Investigating Geography

FOUNDATION

JACKIE ARUNDALE AND GREG HART

Hodder & Stoughton
www.hodderheadline.co.uk

Acknowledgements

The publishers would like to thank the following individuals, institutions and companies for permission to reproduce copyright illustrations in this book:
© actionplus, 3.1tl; © AP Photo/Charles Rex Arbogast, 2.11; © AP Photo/Namas Bhojani, 2.12; © AP Photo/Sherwin Crasto, 3.22; © AP Photo/Eugene Hoshiko, 2.13, 3.21; © AP Photo/Dolores Ochoa, 3.20; AP Photo/Gorans Veljkovic, 3.1br; © Yann Arthus-Bertrand/Corbis, 4.4; Associated Press/AP, 1.21, 2.7, 3.16, 3.17, 3.19, 3.28, 3.29, 5.18a, 6.13; Associated Press, EFE, 1.19; © Steve Austin, Papilio/Corbis, 5.21 (vole); © Paul Barton/Corbis, p4 1.1br; © Jonathan Blair/Corbis, 5.18c; Bluetown History & Arts Centre, 4.2; Joerg Boethling/Still Pictures, 5.6iv; Bright Ideas, 5.29 (logo); Simon Chandler, 5.24, 5.25, 5.26; © Lloyd Cluff/Corbis, 3.27; Corel, p7 (flag), 1.4, 1.13 (flags), p38 (logo), p40br, p52 (flags), p61 (flag), p82 (Cardiff); © Philip James Corwin/Corbis, 6.24; © James Davis, Eye Ubiquitous/Corbis, 5.20c; Cartoon from Thin Black Lines – Political Cartoons and Development Education (DEC), 1.29; Mark Edwards/Still Pictures, 5.4, 5.6ii, 5.18b; Eyewire, p40t; © Friends of the Earth, 6.43; © Colin Garrett, Milepost 92½/Corbis, 6.29; GRAFF/Cartoonists & Writers Syndicate/cartoonweb.com, 2.6; Hodder Arnold, 6.31 (landfill); © Robert Holmes/Corbis, 4.7; © Hulton-Deutsch Collection/Corbis, 1.17; Image Boss, pp16–17 (shadows and bg); Ingram Publishing, p36 (man and computer), p81bg, 5.1, 5.21 (mushroom, fox); Bob Jones, 4.1, 4.3, 4.5, 4.8, 4.13, 4.14, 4.15, 4.18, 4.19, 4.22, 4.23, 4.24, 4.26, 4.27, 4.28, 4.29, 4.30, 4.31, 4.32, 4.34, 4.35, 4.36, 4.37, 4.38, 4.39, 4.40, 4.41, 4.43, 4.44, 4.45, 4.46, 4.47, 4.48, 4.49, 6.2, 6.14, 6.16, 6.17, 6.21, 6.23, 6.25, 6.27, 6.28, 6.31 (all except landfill), 6.32, 6.33, 6.35, 6.36, 6.37, 6.38, 6.39; University Library, Keele/The Warrillow Collection, 6.1; © Frank Lane Picture Agency/Corbis, 5.21 (weasel); Life File/Graham Burns, 3.1tr; Life File/Xavier Catalan, 1.16; Life File/Emma Lee, p4 1.1tr, p5 1.1tl, p5 1.1cr, p5 1.1b, 1.32t; Life File/Barry Mayes, 6.10; Life File/Richard Powers, 6.11; Life File/Eddy Tan, 1.20; J. Freund/Still Pictures, 5.3; © Wolfgang Kaehler/Corbis, 5.20b; © George D. Lepp/Corbis, 5.6iii; © Buddy Mays/Corbis, 5.19, 5.20a; © Amos Nachoum/Corbis, 5.28; Nomad, p7 (tv), pp16–17 (eyes), 3.2b, 4.49bg, 6.9a; Christine Osborne/Ecoscene, 2.18; Photodisc p4 1.1l, p7 (seaside and night), 1.13bg, 1.15bg, 1.32b, 2.1, p35 (monitor), p36 (women), p40cr, p40bl, p40bc, 3.2t, 3.2c, 3.5, 3.15, 3.34bg, 4.6, 4.20, p82 (Mexico), 5.5, 5.6i, 5.8, 5.10, 5.21 (rabbit, insect, small bird, vegetation), 5.23, 5.31, 6.5, p106, p107, 6.41; Gerry Quinn g.quinn@planetquinn.com, 2.9; Adrian Raeside/www.raesidecartoon.com, 2.26; © Roger Ressmeyer/Corbis, 3.11a; © Galen Rowell/Corbis, p5 1.1cl; © Kevin Schafer/Corbis, 3.10a; © Copyright by the local government of Schlema, 6.22, 6.26; © Paul A. Souders/Corbis, 3.4, 5.9; © Studio Burie, Netherlands, 2.2a, 2.2b; © W.T. Sullivan III/Science Photo Library, 2.3; US Geological Survey, 3.1c; © Michael Yamashita/Corbis, 3.26.

(l = left; r = right; t = top; b = bottom; c = centre; bg = background)

The publishers would also like to thank the following for permission to reproduce material in this book: *Guardian* for the adaptation of the extract from 'Thousands left homeless in Peru', 30 June 2001; extract adapted from Insight Guide Spain, ©2002 Apa Publications GmbH & Co Verlag KG (Singapore Branch); UN for various extracts in chapter 2; the extract from the Collins Longman Atlas (redrawn) © Bartholomew Ltd 2002. Reproduced by permission of Harper Collins Publishers www.bartholomewmaps.com. Methuen Publishers for the extract from *Stranger in the Forest* by Eric Hansen (Methuen, 2001); This product includes mapping data licensed from Ordnance Survey® with permission of the Controller of Her Majesty's Stationery Office. © Crown copyright 2002. All rights reserved. Licence number 100019872. Nick Thorpe for the adaptation of the extract from 'Cyanide spill floods into Danube', *Guardian*, 14 February 2000; UN for the UN press release 'Most countries not on track to meet UN's 2015 goals', July 2001; WHO for the Healthy Life Expectancy chart from the World Health Report 2000.

Note about the Internet links in the book. The user should be aware that URLs or web addresses change regularly. Every effort has been made to ensure the accuracy of the URLs provided in this book on going to press. It is inevitable, however, that some will change. It is sometimes possible to find a relocated webpage by just typing in the address of the home page for a website in the URL window of your browser.

Orders: please contact Bookpoint Ltd, 130 Milton Park, Abingdon, Oxon OX14 4SB. Telephone: (44) 01235 827720. Fax: (44) 01235 400454. Lines are open from 9.00 – 6.00, Monday to Saturday, with a 24-hour message answering service. You can also order through our website www.hodderheadline.co.uk.

British Library Cataloguing in Publication Data
A catalogue record for this title is available from the British Library

ISBN 0 340 84632 1

First Published 2003
Impression number   10 9 8 7 6 5 4 3 2 1
Year   2009 2008 2007 2006 2005 2004 2003

Designed by Nomad Graphique.
Typeset by Fakenham Photosetting, Norfolk.
Cover photo from Photodisc.
Printed in Italy for Hodder & Stoughton Educational, a division of Hodder Headline Plc, 338 Euston Road, London NW1 3BH.

Investigating Geography C

Contents

Spain or not?

Figure 1.1

Activities

1 Choose three pictures. For each one complete the following two sentences:
'*I think this photo is probably in Spain because …*'
'*This photo might not be in Spain because …*'

2 Which picture most looks like Spain? Suggest a reason.

3 Which picture does not look like Spain? Suggest a reason.

4 Choose another country. Suggest three pictures to show this country. Suggest why you chose each picture.

ICT Activity

Use the Internet to find pictures for the country you chose for question 4. Use a search engine that has an image search facility, such as **www.google.com**.

What are your first thoughts?

In this chapter we are going to investigate Spain. You are going to investigate whether or not Spain is a **nation**.

You will need to think carefully about the information in this chapter.

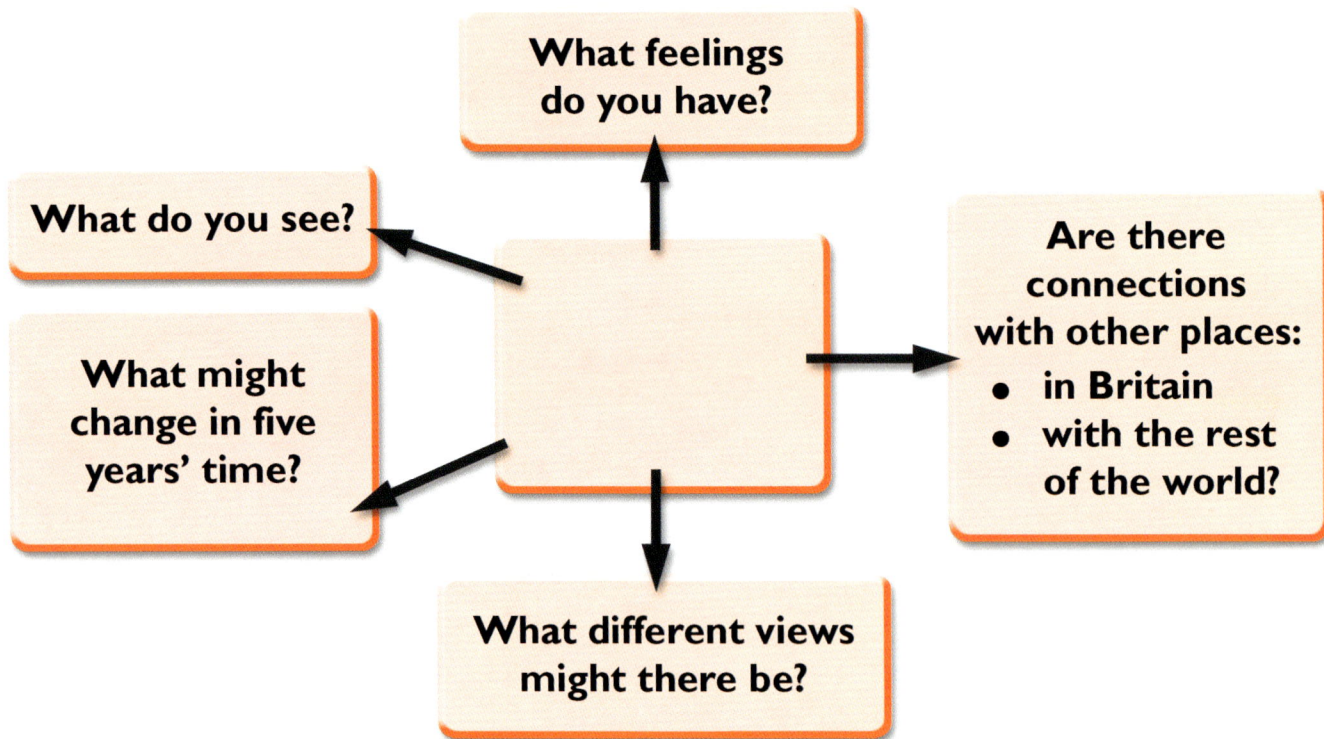

What feelings do you have?

What do you see?

What might change in five years' time?

Are there connections with other places:
- in Britain
- with the rest of the world?

What different views might there be?

△ **Figure 1.2** Learning from an Image frame

Activities

1. Think about what you already know about Spain. Make a list of your ideas.

2. How do you know about Spain? Make a list of all the ways.

3. Choose one of the photographs on pages 4 and 5.

4. Sketch this photo in the centre of an A3 piece of paper.

5. Use the Learning from an Image frame (Figure 1.2) to write some notes around your sketch.

6. Describe your photograph to your class.

How do you develop a viewpoint?

We will never know all there is to know about Spain. We also have to be careful with any information we find because it may be written for a particular purpose. These three stages may be helpful.

Stage A – We are informed

- We get pieces of information from TV, newspapers, teachers and friends about a place.
- These sources will all have a viewpoint that they want us to share.

Stage B – Our experiences

- Our own experience will affect how we feel about the information we have received.
- New information can change what we know.

Stage C – Developing knowledge

As we grow up and learn more, we might change how we felt about things earlier.

△ **Figure 1.3** Sense of Place frame

Activities

7 Imagine your friend has had a bad holiday in Spain. Write a paragraph to describe the holiday.

8 How might your friend's story affect your view of Spain?

9 Read Figure 1.3. Why do you have to be careful with stories like the one your friend told you?

What do you know about Spain?

Activities

1 You and a partner are going to draw a map of Spain. You will study Figure 1.5 for five minutes. Then close the book. Draw the map from what you can remember onto a blank outline. Before you start, talk to your partner about the best way to do this task.

2 Check your completed map with Figure 1.5. Add any important missing detail. How well did you do?

3 Compare your map with others' maps. Give each map a mark out of 10. Tell the others what was good about their map. Suggest what else they could have included.

4 Find maps of Spain in an atlas. Add 10 other pieces of information to your map.

5 Add some of the facts from page 9 to your map.

6 Write five sentences about Spain.

Useful words

These phrases may be useful:

English	Spanish
hello	hola
goodbye	adiós
good morning	buenos días
good afternoon	buenas tardes

English	Spanish
yes/no	si/no
please	por favor
thank you (very much)	(muchas) gracias
where is...?	¿donde está...?

These phrases may be useful:

English	Spanish
excuse me, I don't understand	perdón, no entiendo
could you please speak more slowly?	¿podría hablar más despacio, por favor?

English	Spanish
I am/my name is...	soy/me llamo...
I am from...	soy de...
I am English	soy inglés
what is your name?	¿como se llama usted?

English	Spanish
how are you?	¿como está?
very well	muy bien

△ **Figure 1.4** Useful Spanish words and phrases

Spain makes up 85% of the Iberian **peninsular**. 88% of its boundary is with water. France is to the north and Portugal to the west. The small British colony of Gibraltar is on the southern tip of Spain. The Balearic Islands in the Mediterranean and the Canary Islands off the coast of Africa are also part of Spain.

Spain has a population of just over 40 million (2001). 77% of the population live in towns and cities.

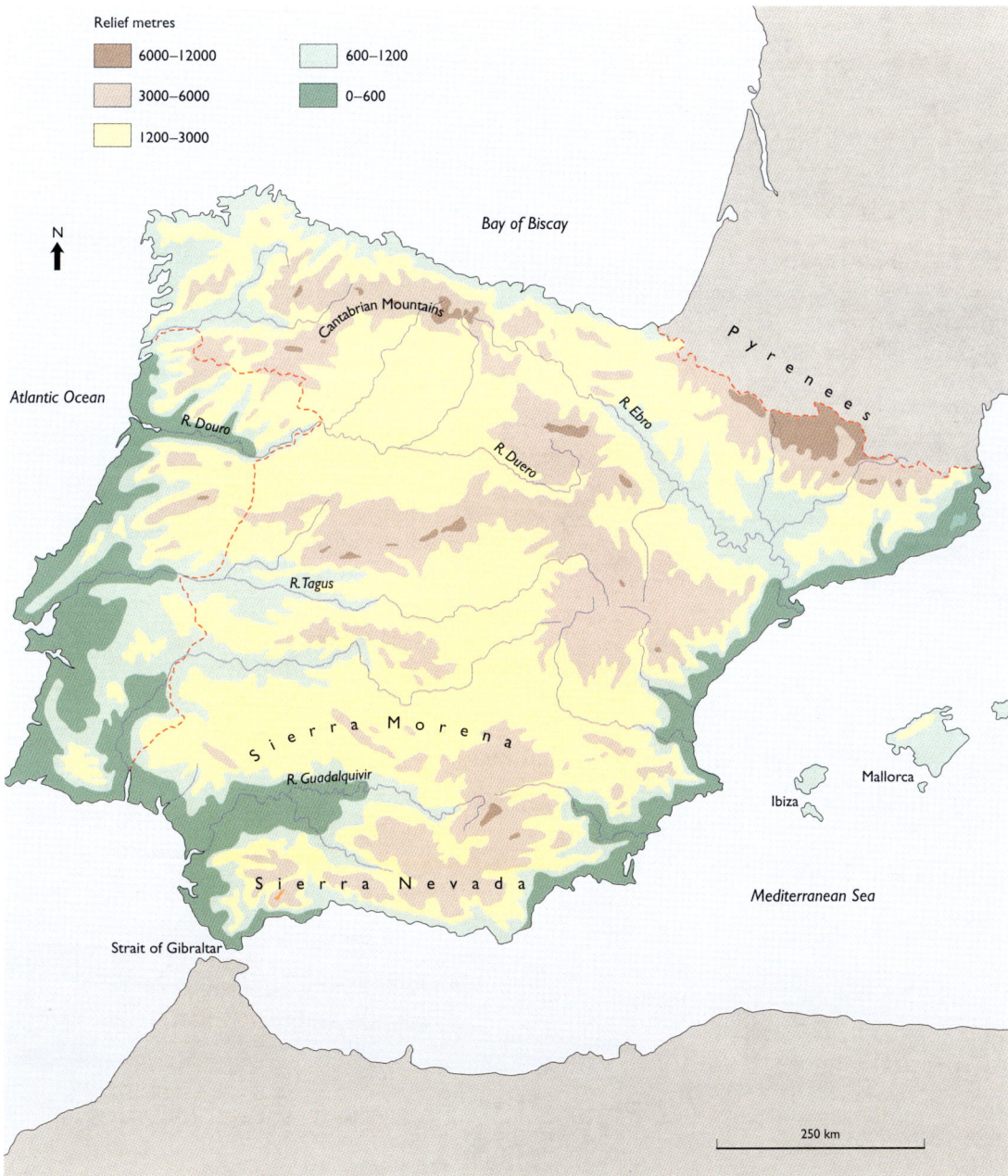

◁ **Figure 1.5** Relief map of Spain redrawn from © Bartholomew Ltd 2002. Reproduced by permission of Harper Collins Publishers www.bartholomew maps.com

How to make sense of Spain?

So far in this chapter, we have used words, pictures and maps to find out about Spain. We are now going to investigate some statistics about Spain's population. We will compare Spain with four other countries. In 2001 Norway was top in the UN's list of best countries to live in.

Country / year	1975	1999	2015
Norway	4	4	5
United States	220	280	321
Japan	112	127	128
United Kingdom	56	59	61
Spain	36	40	39

△ **Figure 1.6** Total population – millions

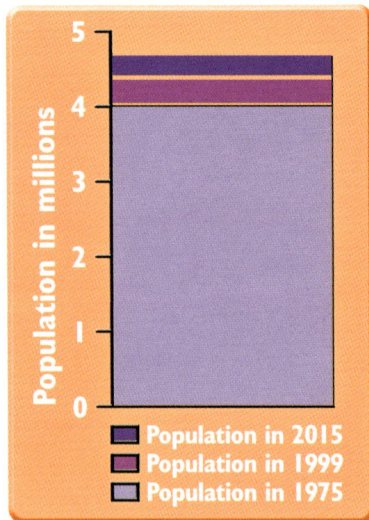

◁ **Figure 1.7** Norway: bar graph showing population in millions, 1975–2015

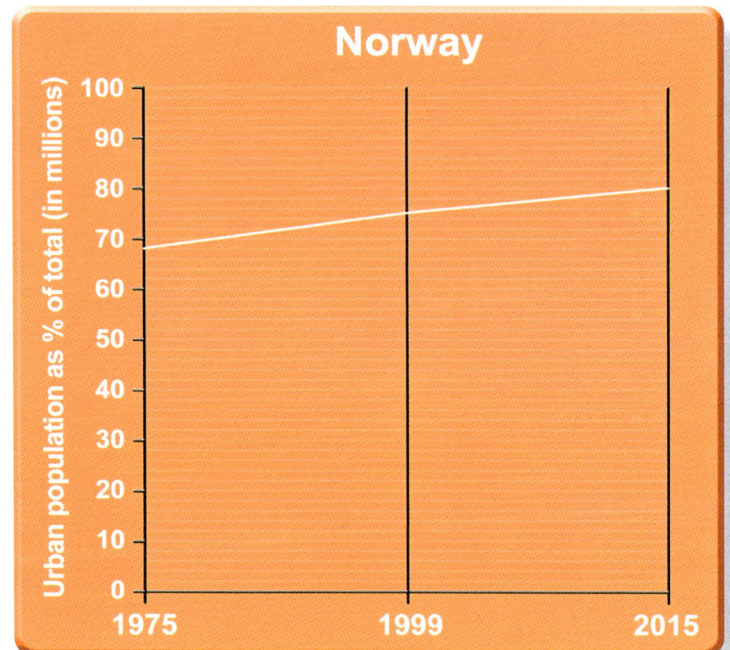

△ **Figure 1.9** Line graph to show Norway's urban population 1975–2015

Country / year	1975	1999	2015
Norway	68	75	80
United States	74	77	81
Japan	76	79	82
United Kingdom	89	89	91
Spain	70	77	81

△ **Figure 1.8** Urban population – as a percentage of total population

Country / year	1999	2015
Norway	20	16
United States	22	19
Japan	15	13
United Kingdom	19	15
Spain	15	13

△ **Figure 1.10** Population aged 15 and under – as a percentage of total

Country / year	1999	2015
Norway	16	18
United States	12	14
Japan	17	26
United Kingdom	16	19
Spain	17	20

△ **Figure 1.11** Population aged 65 and above – as a percentage of total

Country / year	1999	2015
Norway		
United States		
Japan		
United Kingdom		
Spain		

△ **Figure 1.12** Population aged between 15 and 65 – as a percentage of total

Activities

1 Figure 1.7 shows a divided bar graph for Norway. On a blank copy of these axes draw a divided bar graph for the UK using the data in Figure 1.6.

2 Study Figure 1.6. Suggest why drawing a divided bar graph for Spain might be difficult. Can you suggest how the graph for Spain could be drawn?

3 Study Figure 1.6 and use an atlas. Which country
• is the largest
• is the smallest
• has a rapid population growth
• is the most crowded?
How do you know?

4 Copy Figure 1.9. Copy the line that shows Norway's urban population growth. Use the data from Figure 1.8 to plot the USA, Japan, the UK and Spain. Use a different colour for each country.

5 State for each country if their urban population:
• has stopped growing
• is growing slowly
• is growing quickly.

6 Suggest what might happen to Spain's cities by 2015.

7 Figure 1.10 shows the percentage of people under 15. Figure 1.11 shows the percentage of people over 65. By adding these two numbers, we get the **dependant population**. Take this combined number away from 100% and you find out the percentage who are working. For example:

In 1999, Norway has 20% under 15 years and 16% over 65 years. The dependant population is 36%. The percentage working is 100% − 36% = 64%.

Copy Figure 1.12. Work out the percentage of dependants in each country for 1999 and 2015.

8 Write five sentences about Spain's population using the information on these two pages. Use these prompts:
• Spain's population is …
• The number of people living in cities is …
• The percentage under 15 years is …
• The percentage over 65 years is …
• By 2015 the percentage over 65 years will …

One Spain or several?

Las Españas

△ **Figure 1.13** Spanish national flag and flags of the regions

Spaniards often speak of their country as Las Españas (plural), because they feel that there is more than one Spain. Spain is one of the largest countries in Europe. It has a wide variety of climates and landscapes. It also has a mixture of people. There is no such thing as 'typically Spanish'.

The populations in several regions of Spain have kept a separate identity. There are several different languages. The regions include the Basques, about 2.1 million people who live mainly around the Bay of Biscay; the Galicians, about 2.5 million people in the north-west of Spain; and the Catalans of eastern and north-eastern Spain.

Activities

1 On an outline map of Spain show the three different regions mentioned in the paragraph opposite. Check your answer in an atlas.

Places or spaces?

We are now going to look deeper into how politics and power in Spain have affected where people live and how they feel about that place.

What are you?

If asked 'What are you?' most people will answer, 'I am English or American or French…' and then proceed to specify, 'I come from Devon or Florida or Normandy…' ending up with 'My home town is Exeter or Miami or Cherbourg…'. The Spaniard, on the other hand, will reverse this order and start with 'I am from Denia' (most important), 'I am Valencian' (of secondary interest) and, if pressed further, will admit 'Oh yes, I am Spanish'. The Spaniard's **individualism** is reflected in this peculiar identification with the local area rather than the country as a whole. Anyone coming from outside Spain is an *extranjero* but even a Spaniard from the next village will be referred to as a *forastero* (stranger). To this day, Spaniards retain this extreme individualism.

△ **Figure 1.14** An extract from *Culture Shock: A Guide to Customs and Etiquette, Spain* by Marie Louis Graff

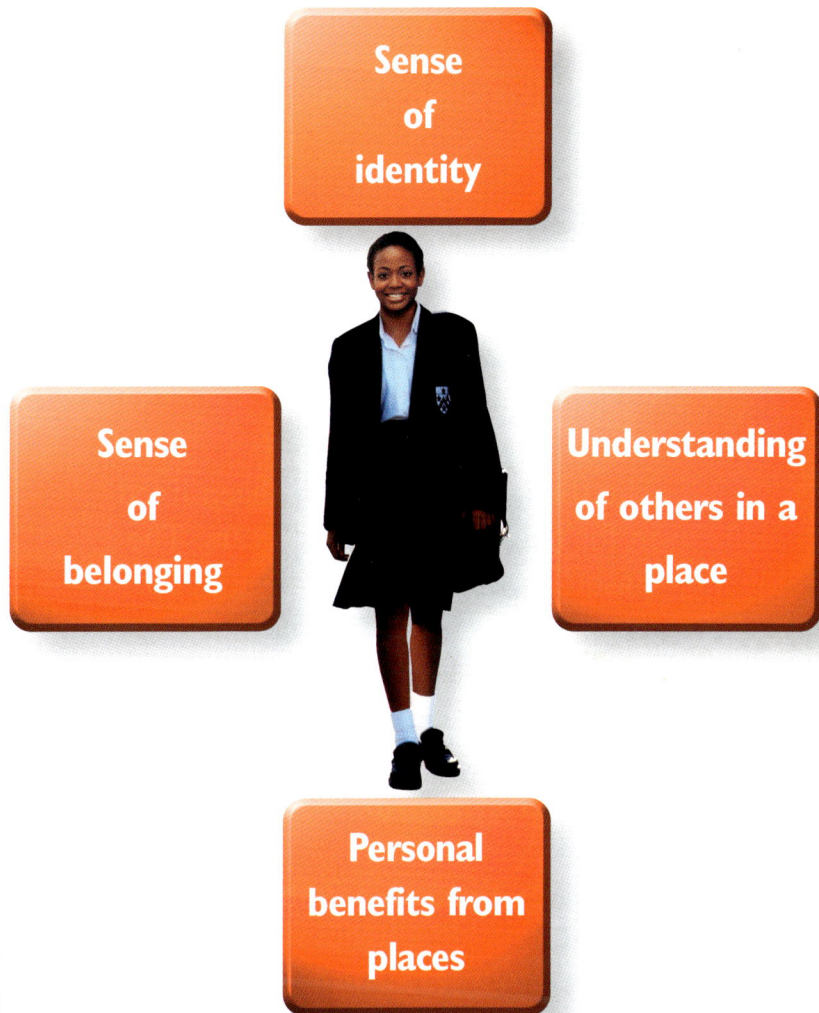

Sense of identity

Sense of belonging

Understanding of others in a place

Personal benefits from places

△ **Figure 1.15** Sense of Belonging diagram

Activities

2 Your teacher will read Figure 1.14 to you. If there are any words you do not understand, ask your teacher. What is this passage about?

3 Your teacher is going to read the passage again. While you listen, draw the ideas as pictures. These pictures represent the feelings in the passage.

4 Make a copy of the diagram above (Figure 1.15). Add notes about the place that you feel that you belong to.

Spain – geography or history?

To explain Spain's geography we need to know a little of its history. It is important to understand why certain people feel such a sense of belonging to certain places within Spain.

A brief history of the people in Spain

The Iberian peninsula has been lived in for many thousands of years. The Basque people were the first group. The Romans conquered it in the first century BC. They kept control until European barbarian tribes and then Muslims (Moors) from North Africa took over in 711 AD.

Gradually the native Catholics fought back. In 1492 one ruling family took over. During the 16th century, Spain became the most powerful nation in Europe. Spain took over other countries, particularly in South America. However, Spain lost much of this empire in the 17th century. Spain became less and less powerful into the 20th century.

▷ **Figure 1.16** A Spanish galleon active in the 16th century, now a tourist attraction in Barcelona Maritime Museum

Spain's modern history

Spain was not directly involved in either the First or the Second World Wars but, in 1936, **civil war** broke out. This was caused by fears that there would be a socialist revolution in Spain. The Nationalists and General Franco won. Spain became an isolated country and did not join the United Nations until 1955. Franco was a **dictator**; he did not share power. Some of the **regions**, for example the Basque country, hated this. During the 1960s and 1970s, Spain became a modern country with a growing tourist industry. Franco died in 1975 and was succeeded by King Juan Carlos. In 1977 Spain held its first elections since 1936 and is now a **democracy**. Each of its 17 regions rules itself; very different to Franco's rule.

What has Spain's history got to do with its geography?

The Basque country, Catalonia and Galicia consider themselves to be historic nationalities. Before Franco they had been independent from central government since the 16th century. In these regions local languages are spoken.

△ **Figure 1.17** Men in Madrid clear rubble following bombing by Nationalist rebels

Activities

1 Work in a pair. Draw a timeline and mark on all the important events in Spanish history that are mentioned.

2 Why does Spain's history mean that Spain is several nations?

Is Basque terrorism Spain's biggest problem?

The issue of the Basque People – a problem for Spain

About 75% of Spaniards think that Basque terrorism is the biggest problem for Spain. In recent years there have been car bombs and assassinations. In Barcelona, over 900 000 people took part in a march to protest about the violence. The Spanish Prime Minister insists they are not going to allow Basque terrorists to force terror on Spain.

The Key Issues

- Who is a Basque?
- Where is the Basque territory?

Spain says three provinces are 'Basque country' but the people who want a separate country also want another province and part of Southern France. They want this as a homeland for 3 million Basque people.

ETA, the Basque terrorist organisation, have been blamed for 800 deaths since 1968. Many respected world leaders have asked for the violence to end but there are still bombs and fighting.

Basque identity

Basque Nationalism began about 100 years ago when immigrants arrived in the area looking for work in the factories. ETA started killing and fighting when General Franco was trying to squash the Basques and wipe out their language.

Franco died in 1975 and now the Basques have a lot of say, running their own schools, parliament and police force. ETA still want more independence and the violence continues.

People thought the violence might end when ETA called a ceasefire in 1998, but a year later a car bomb in Madrid started it all over again. Most people want a peaceful solution to the issue and a final end to the violence and fear.

△ **Figure 1.18** Information on the Basque question, researched from the Internet

◁ **Figure 1.19** Car bombs have been part of ETA's fight for Basque independence

▽ **Figure 1.20** The Basque landscape

▽ **Figure 1.21** Demonstrations against ETA violence

Activities

1 Read the Internet article from page 16 together with your class. Then discuss the following questions.
  a What is the disagreement about?
  b Who are ETA and what have they done?
  c What changes have there been since Franco died?

2 Look at the photos carefully. For each one complete the following sentences to explain what the image tells us about the Basques' region or conflict.
This photo shows... It tells me that...

One nation?

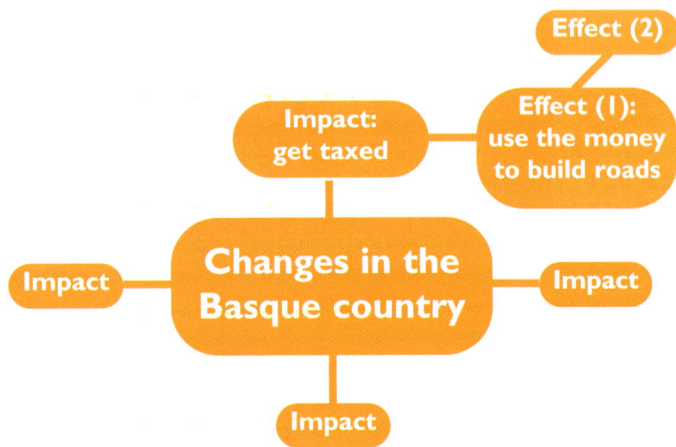

△ **Figure 1.22** Effects Wheel

Five questions to ask

1. **IMAGES** – What do you, and others, think this conflict is really about?
2. **BACKGROUND** – What has actually happened so far, and why?
3. **SOLUTIONS** – What are all the possible solutions that you can think of?
4. **CHOICES** – What are the best possible solutions?
5. **ACTION** – What, if anything, can you do about this issue?

IMAGES
BACKGROUND → CONFLICT → SOLUTIONS → CHOICES → ACTION

△ **Figure 1.23** Five Questions to Ask frame

Activities

1 Copy Figure 1.22. In a pair, note down four impacts if the Basques were to get their own country.

2 Ask other pairs for their ideas. You can add any ideas to your chart.

3 For each impact, suggest some effects on the Basques. Add these to your chart.

4 Study Figure 1.23. This is a way to think through the Basque problem. Draw out the flowchart. In each box write some notes.

ICT Activities

There are two other regions that would like to be independent. These are the regions of Catalonia and Galicia. The maps opposite locate these regions.

1 Use an Internet search engine to find out about one of these regions.

2 Use Publisher to design a leaflet on this region.
Include this information:
• a map
• pictures
• some facts
• reasons why they want to be independent.

◁ **Figure 1.24** The regions of Spain

▽ **Figure 1.25** Galicia

△ **Figure 1.26** Catalonia

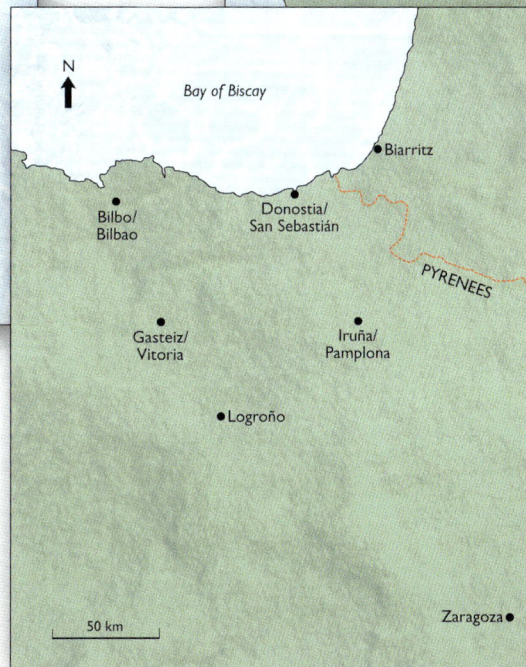

▷ **Figure 1.27** Basque country

Assessment tasks

Is this Spain?

A Spanish person's view

'My **ancestors** come from Galicia. This is one of the most influential parts of Spain. Spain should be one, politically, economically and industrially. Spain needs to be part of the European Union. I believe that everyone should be able to see themselves as Spaniards. They should work together to make Spain a better place. Not everyone agrees with me. Some areas believe they are different. They have different identities. They speak a different language and have different ways. The three areas where this happens are: Catalonia, Basque country and Galicia.

Although these regions have differences, they should be loyal to Spain. Our language and our nationality is Spanish. These are the major reasons for Spanish citizens to unite.

Catalonia, the Basque country, and Galicia are not strong enough to survive by themselves. Catalonia needs the rest of Spain to support it. The same is true for the Basque country and Galicia. 'The Union of Spain makes the power' is a popular Spanish saying. It should be used in real life.

△ **Figure 1.29** Working together?

△ **Figure 1.28** Adapted from an extract from a Spanish person's own website. The full text can be found at **http://dons.usfca.edu.traned2/spain.htm**

Target task

1 Listen to your teacher read Figure 1.28.

2 Does the writer want a united Spain?

3 Copy Figure 1.30.

△ **Figure 1.30** How? and Why? frame

4 Find the key points in Figure 1.28 that support your answer.

5 Write these on your How? and Why? frame (Figure 1.30).

6 Copy Figure 1.29 in the middle of a piece of paper. What do you think each row means for Spain? Write this on your paper.

7 Use the writing frame (Figure 1.31) to explain the two views about Spain and the regions.

The issue that we are thinking about is the Basque region in Spain.

Some people think that ....
because ....
They argue that ....
because ....
On the other hand .... disagree with the idea that ....
They claim that ....
They also say that ....
My opinion is ....
because ....

△ **Figure 1.31** Writing frame

△ **Figure 1.32** What is the real Spain?

Review

What is the 'Real Spain'? A gypsy caravan towed along an autopista [motorway] behind a gleaming BMW? Pale-skinned foreigners stretched out in neat rows on some sandy beach, broiling themselves a deep shade of lobster? The 'Real Spain' is an elusive concept.

◁ **Figure 1.33** Adapted from *Culture Shock: A Guide to Customs and Etiquette, Spain* by Marie Louis Graff

Activities

1 Look back at the images on pages 4 and 5. All of these are photos of Spain. These photos are very varied, as is Spain itself.
a Make a list of seven different photos or images that you would now choose to show someone to give them a glimpse of Spain. Draw a copy of the table below to record your choices in.
b Explain why you would choose each photo.

2 Look back at your work on Spain. Write sentences to describe:
• One thing that has surprised you
• One area of Spain you would like to visit and why

• Changes that might take place in Spain in the next 20 years.

3 Look at the quote (Figure 1.33). How would you reply to the question, 'What is the real Spain?'

4 Think about another country that you would like to investigate. What would you need to look at to build up a proper understanding of that place?

5 Work as a group to produce a series of questions that you would need to answer in order to really understand what a country is like.

Photo number	What it shows	Why you have chosen it
1		
2		
3		
4		
5		
6		
7		

2 Development: Global scale

What can maps show?

This chapter is about **global** development issues. It will explore ideas about development at different scales. You will think about development at a personal, local, regional, national and international scale. Different types of information will help you understand **patterns** and **distributions** of development.

△ **Figure 2.1** The Earth from space

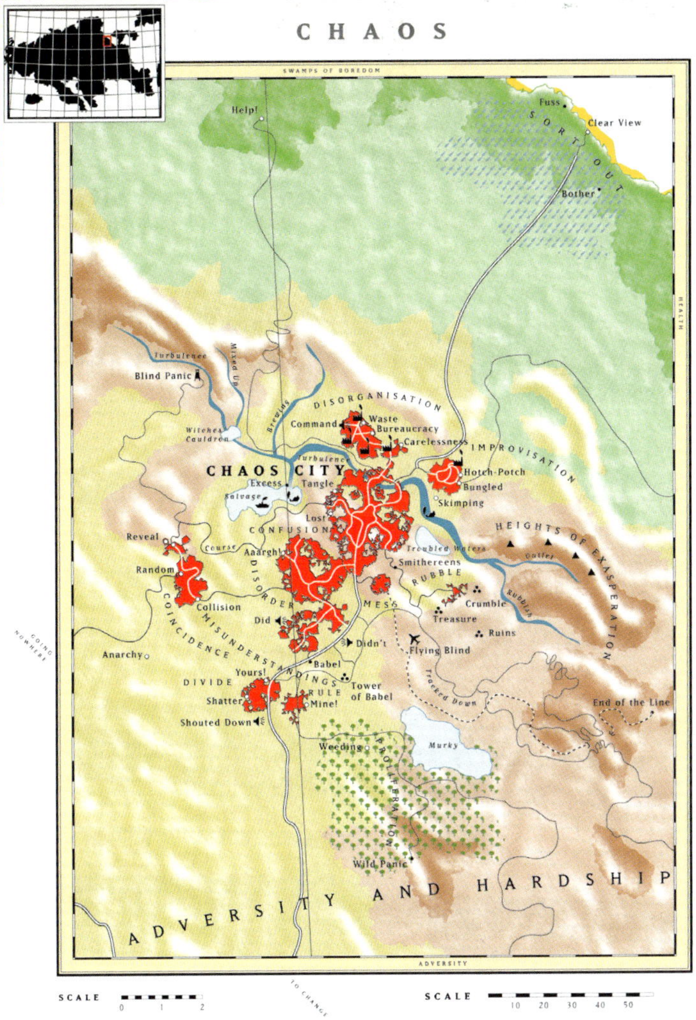

△ **Figure 2.2a** Map of Chaos from *The Atlas of Experience*

△ **Figure 2.2b** Key for Map of Chaos

Activities

Figure 2.2a is an unusual map. It comes from *The Atlas of Experience* produced by **cartographers** Louise Van Swaaij and Jean Klare. The map shows rivers, mountains and towns. At first, the map looks like a far-away place. If you look closely, you will see that the map is 'surveying our shared world of thoughts and emotions' (Van Swaaij and Klare). The map is a mix of reality and fantasy.

1 With a partner study Figure 2.2a.
 a List ten place names.
b Use a dictionary to find what each word means.
c Make up a class list.

d Decide whether each word is a feeling or a name.
e Suggest why the cartographers used these words on their map.
f What do you think/feel about this map?

2 Draw your own map of experience. Use the legend from Figure 2.2b. You could either make up a place or use where you live. Try to map how you feel about this place. Choose a title for your map.

3 Display your class' completed maps. These maps will be useful for other tasks. Spend time looking at the maps and discuss what they tell you about our views of places.

Do maps tell the truth?

'Geography' means 'writing about the earth'. Your map is a way of writing about your world. There are other maps to show the world. These maps are not always the same as reality. Some maps select the information they use to give a biased view of the world. It is important to look carefully at maps. Figures 2.3, 2.4 and 2.5 show aspects of global development.

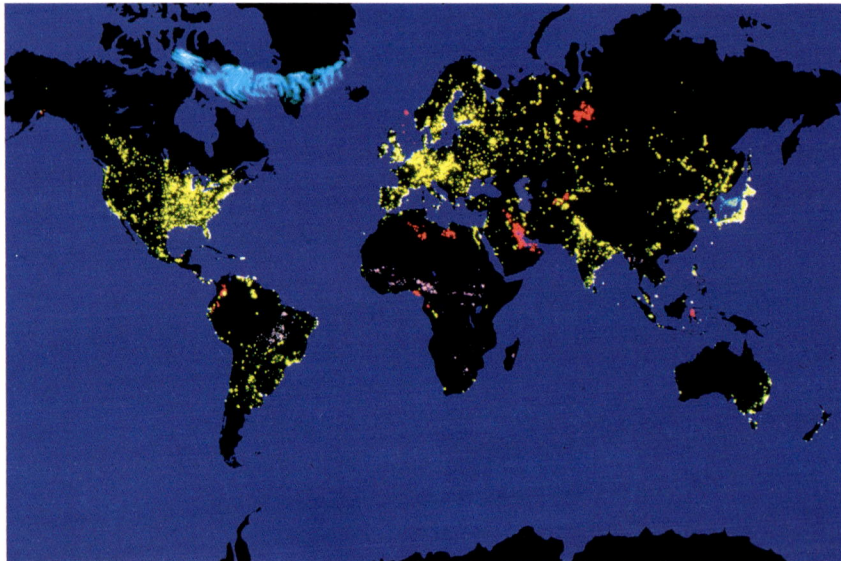

◁ **Figure 2.3** The Earth at night. This is a satellite-computer image. Yellow lights show urban areas; red are oil well flares burning; purple are vegetation; green are boats fishing for squid; and blue is the aurora borealis

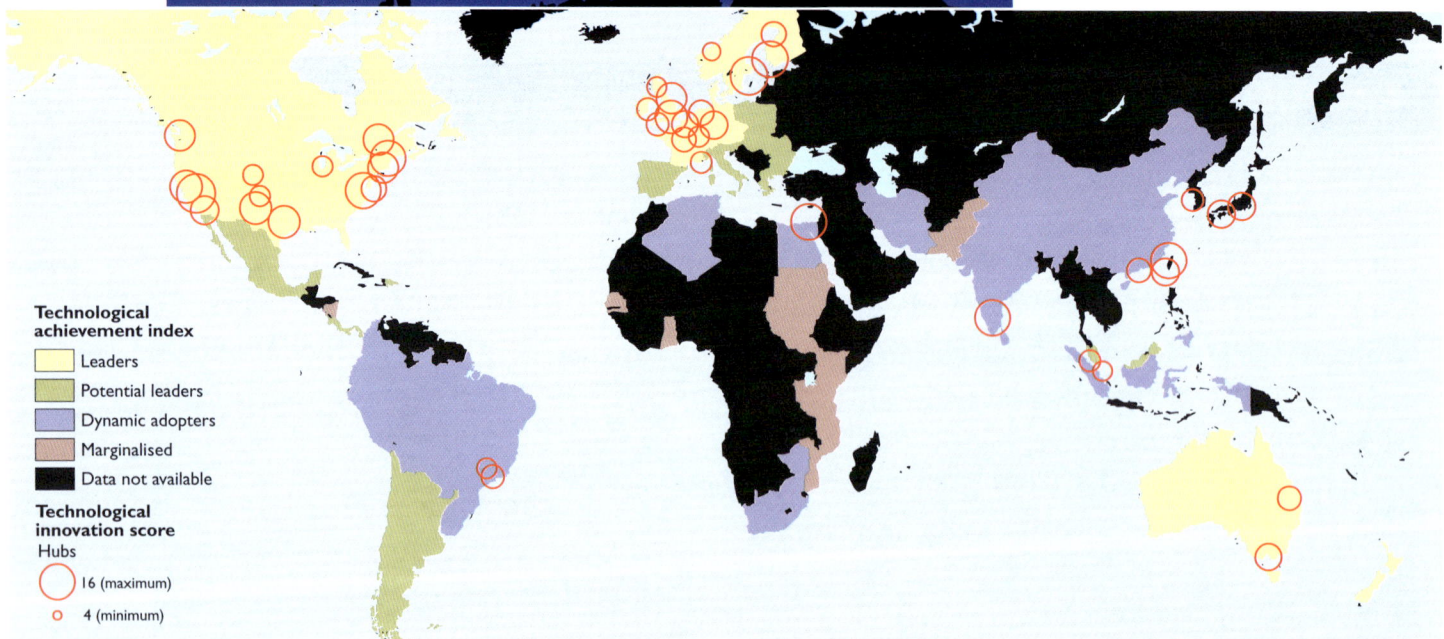

Technological achievement index
- Leaders
- Potential leaders
- Dynamic adopters
- Marginalised
- Data not available

Technological innovation score
Hubs
- 16 (maximum)
- 4 (minimum)

△ **Figure 2.4** The Geography of **Technology** map. This map is based on a survey by *Wired* magazine to find the most important centres for digital geography. The survey found 46 centres. It also grouped countries by their use of this technology. The leaders invented ideas and some actively adopted these ideas. These countries had important centres or hubs. Others lacked the money and skills to use this technology

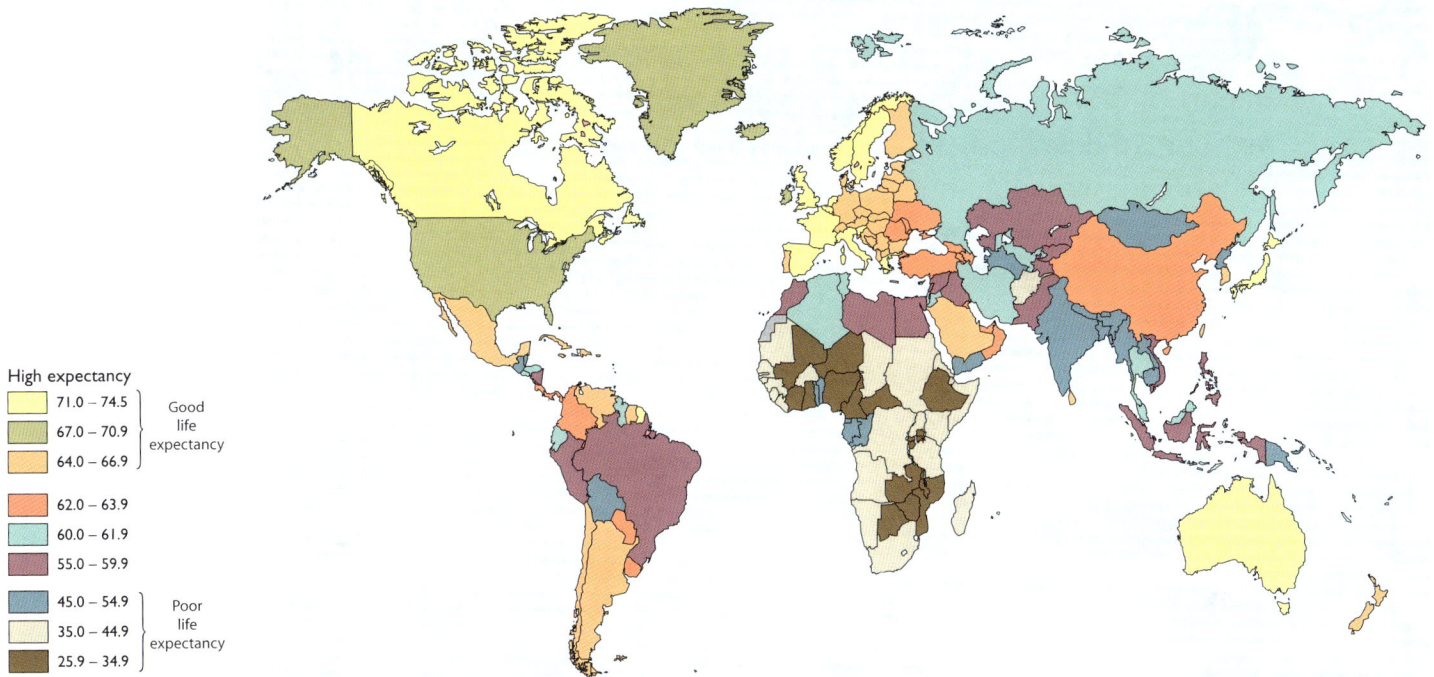

High expectancy

71.0 – 74.5	Good life expectancy
67.0 – 70.9	
64.0 – 66.9	

62.0 – 63.9
60.0 – 61.9
55.0 – 59.9

45.0 – 54.9	Poor life expectancy
35.0 – 44.9	
25.9 – 34.9	

Low expectancy

No data

Life expectancy means the average life span in years.

Measure: Disability adjusted life expectancy at birth, both sexes, estimates for 1997

The boundaries and names shown and the designations used on this map do not imply the expression of any opinion whatsoever on the part of the World Health Organization concering the legal status of any country, territory, city or area or of its authorities, or concering the delimitation of its frontiers or boundaries. Dotted lines on maps represent approximate border lines for which there may not yet be full agreement.   © WHO 2002. All rights reserved

△ **Figure 2.5** Healthy **life expectancy** from the World Health Report, 2000

Activities

Study Figures 2.3, 2.4 and 2.5. These show different views of global development. With a partner, choose one of these maps to study. You will need an atlas that shows countries. Complete the following sentences about your chosen map.

The map we chose to study is ....
This map shows ....
This is a useful map because ....
The map could be improved by ....
One difference between Africa and North America is ....
Another difference is ....

The most developed continent is .... because ....
The least developed continent is .... because ....
From the other two maps the most developed continent is .... because ....
The least developed continent is .... because ....
These maps do show a true view of global development because

....

Or

These maps do not show a true view of global development because

....

Whose choice?

The Human Development Report of 2001 suggested that human development is not about how much money you have. It is about helping people to be creative and have a worthwhile life, to meet their needs and interests. 'People are the real wealth of nations.' Development is providing choice for people to lead lives that they value.

Activities

1 Study Figure 2.6.
a What do you think the pig might be saying?
b Who is the 'pig' supposed to be?
c Write down three things the pig might say about development at the global scale?
d What ideas do others have in your class?
e Choose your best idea and add this to the cartoon.

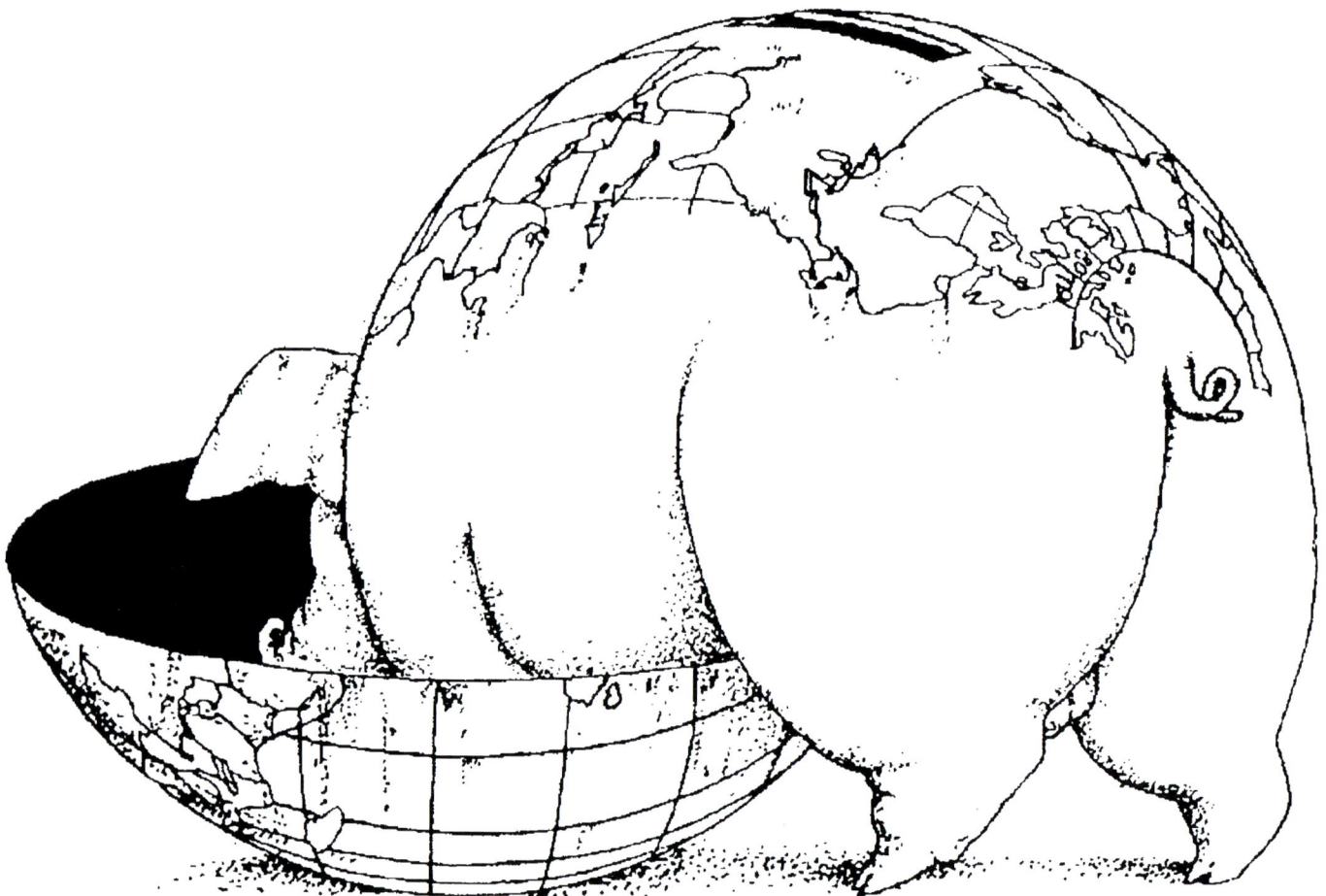

△ **Figure 2.6** Global connections?

Development patterns

Activities

Look back at Figures 2.3, 2.4 and 2.5 (pages 26–7). In an atlas find maps that show measures of development, for example, employment and car ownership.

2 On a copy of the map below, write 10 ideas about global development using Figures 2.3, 2.4 and 2.5.

3 Which is the most developed part of the world?

4 Read the caption for Figure 2.7. What do you think that Kofi Annan means?

5 Describe how talking to different people helps people to understand other places.

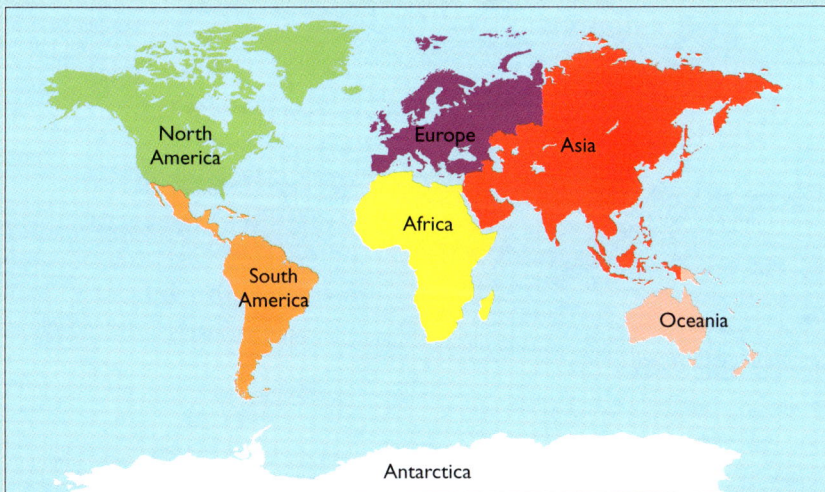

△ Outline map of the world

▷ **Figure 2.7 United Nations** Secretary General Kofi Annan, 2001. He believes that by talking, different groups of people can get to know each other better

Do you know that . . .

...emailing a 40-page document from Chile to Kenya costs less than 10 cents, faxing it about $10, and sending it by courier $50?

...in 2001 more information can be sent over a single cable in a second than in 1997 was sent over the entire Internet in a month?

...the cost of transmitting a trillion bits of information from Boston to Los Angeles has fallen from $150 000 in 1970 to 12 cents today?

...there are expected to be 1 billion users of the Internet in 2005?

...in two years from 1998 to 2000 Internet users increased from 1.7 million to 9.8 million in Brazil, from 3.8 million to 16.9 million in China and from 2 500 to 25 000 in Uganda?

...Cuba developed the only vaccine against meningitis B, through biotechnical research providing national immunisation by the late 1980s?

...India's information and communication technology exports rose from $150 million in 1990 to nearly $4 billion in 1999?

...between 1992 and 1997 Vietnam reduced the death toll from malaria by 97% and the number of cases by almost 60% by developing and using locally produced, high-quality drugs?

...in Brazil a team of computer scientists, commissioned by the government, have designed a basic computer for around $300?

...pharmaceutical sales in Africa are forecast to be just 1.3% of the global market in 2002?

...just 0.1% of the 25 million in sub Saharan Africa have access to HIV/AIDS drugs?

Note: You may want to use the Internet to find out the current dollar exchange rate.

Technology networks are transforming the traditional map of development, expanding people's horizons and creating the potential to realize in a decade progress that required generations in the past

HUMAN DEVELOPMENT REPORT 2001

MAKING NEW TECHNOLOGIES WORK FOR HUMAN DEVELOPMENT

△ **Figure 2.9** Each year the United Nations publishes a Human Development Report (www.undp.org/hdro). This looks at major global issues

Activities

1 Read the fact bubbles on page 30.

a Copy the five fact bubbles that are the most surprising.

b Put these in order of the most to the least surprising.

c Which are the same as others in your class? Which are different from others in your class?

d Which ones are about health? Which are about computers? Which are about communication?

2 **a** Label on a blank world map each of the countries mentioned in the fact bubbles.

b Use one or two of these symbols to show the fact on the map:
  • C = communications
  • I = Internet/computers
  • D = use of medicine

c Make sure that your map has a title, key and scale.

3 Suggest how technology will help many poorer countries in the future.

What is the role of technology in development?

The Human Development Report 2001 argues that new technologies can help reduce world poverty.

It also suggests that technology is not a luxury for rich countries.

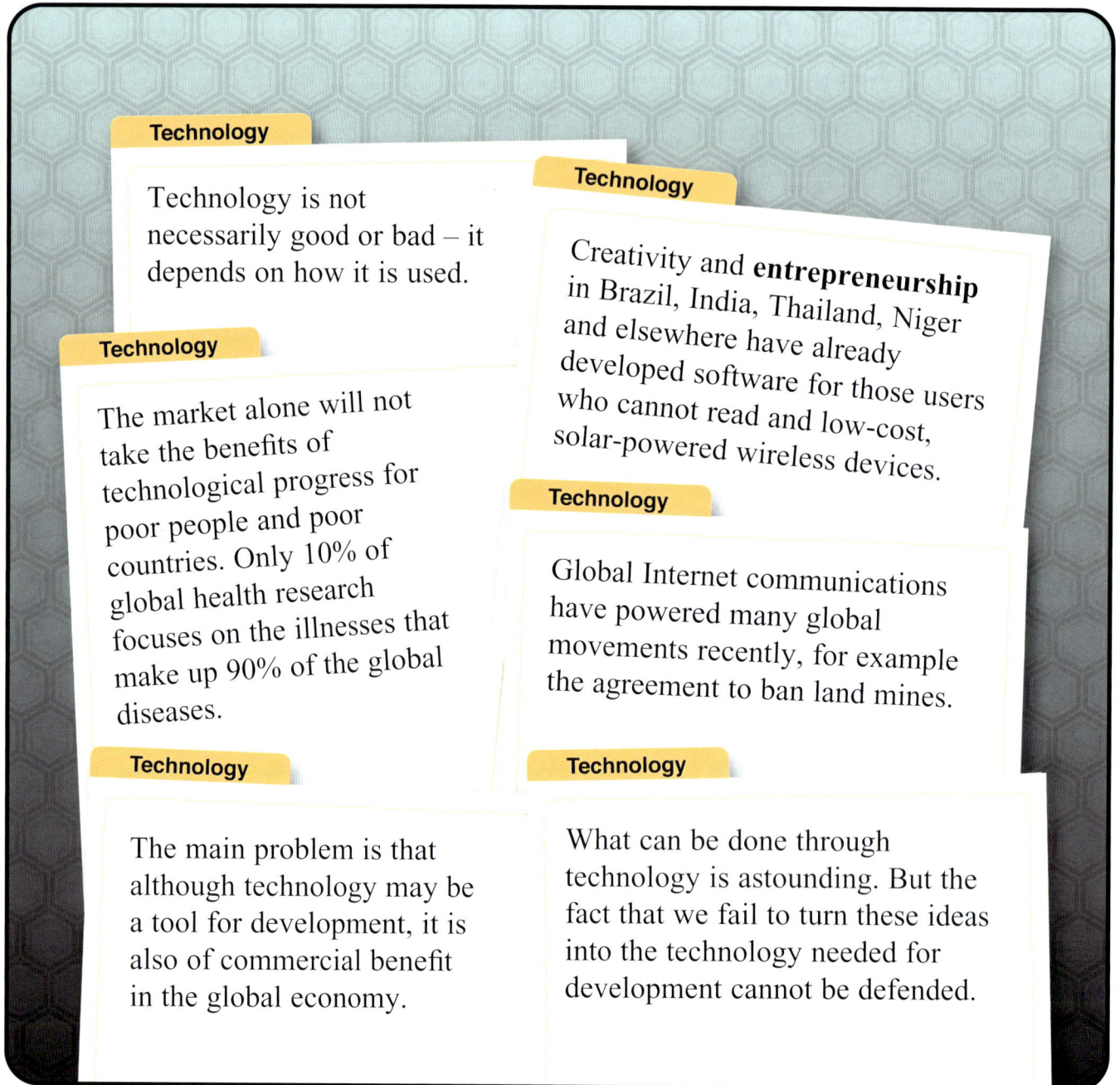

Technology

Technology is not necessarily good or bad – it depends on how it is used.

Technology

The market alone will not take the benefits of technological progress for poor people and poor countries. Only 10% of global health research focuses on the illnesses that make up 90% of the global diseases.

Technology

The main problem is that although technology may be a tool for development, it is also of commercial benefit in the global economy.

Technology

Creativity and **entrepreneurship** in Brazil, India, Thailand, Niger and elsewhere have already developed software for those users who cannot read and low-cost, solar-powered wireless devices.

Technology

Global Internet communications have powered many global movements recently, for example the agreement to ban land mines.

Technology

What can be done through technology is astounding. But the fact that we fail to turn these ideas into the technology needed for development cannot be defended.

△ **Figure 2.10** Quotes adapted from Human Development Report, 2001

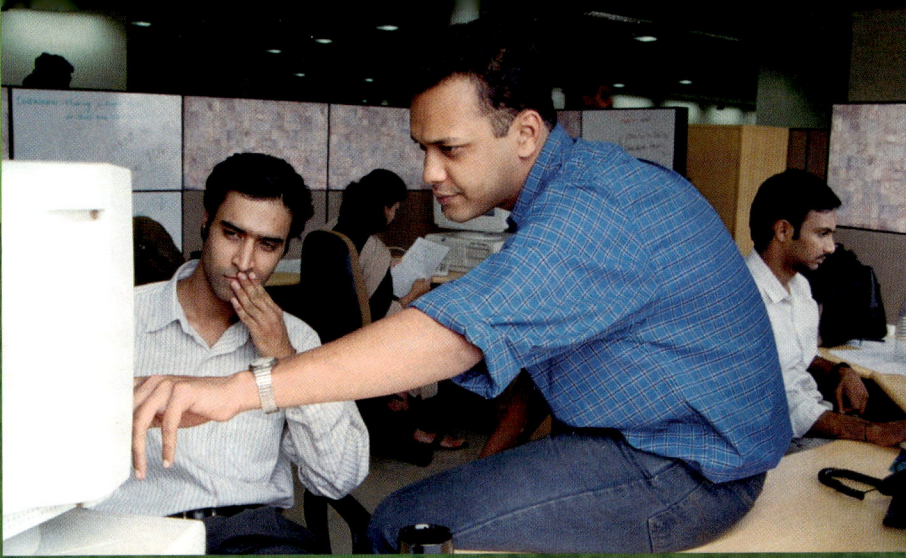

△ **Figure 2.11** Computer training in India – all countries should train their people to use the new technology

△ **Figure 2.12** Kosovan refugees use the Internet as part of their education and to access information

Case Study 1

Open source software is a low cost idea that can be shared. It is made up of contributions from around the world. The software details are open for all to use. The low cost means government ICT budgets go further. Using open source software could help the global ICT revolution take off.

Case Study 2

In Sri Lanka, the Koyhmale Community Radio Station helps its listeners in remote areas to use the Internet. Schools send requests for information. The radio station searches the Internet and broadcasts the information. It also mails this to the school or through the station's open resource centre. The station uses local languages rather than English.

△ Case studies, adapted from the Human Development Report, 2001

◁ **Figure 2.13** A satellite dish in Afghanistan. Making these locally provides jobs and cheaper dishes. More people can have access to broadcasts

What's your view?

Use the information on pages 32–3 to give a talk about Development and Technology.
The presentation will be in the form of a 'SOFT' report: Strengths, Opportunities, Failures and Threats to technology-aiding development.

1 In a group of three, work on the SOFT report. Use a copy of the frame below to organise your thoughts.

Talk title: SOFT report on Development and Technology	
Strengths (good things about using technology) • • •	**Opportunities (how it will help people)** • • •
Failures (problems at the moment) • • •	**Threats (what could stop the use of technology)** • • •

△ **Figure 2.14** SOFT report frame

2 Share with the rest of your class your SOFT ideas. You can add other ideas.

3 Write a list or brainstorm ideas about the important key words and ideas.

Activities

4 The talk should last two minutes. Each of your group should say something. Use the Thinking frame to draft your talk. Everyone will need their own rough copy of this Thinking frame.

Part of your talk	What you are going to say	Who is going to say it
Main ideas		
What are the strengths?		
What are the opportunities?		
What are the failures?		
What are the threats?		
What is the big point you want to make?		
Summing up Is technology the way to improve people's lives? Can technology make a difference?		

△ **Figure 2.15** Oral presentation Thinking frame

5 During the planning of your talk, choose one of your group to talk with the other groups. Bring back any ideas or different points of view that might help your group.

6 Each group will judge the talks. As a class, decide on five criteria for judging the talks. Maybe score each out of five and give an overall total.

7 When every group has spoken, each group should decide which was the best talk and some reasons why. Write this up and share it with the class.

8 Think about how you worked as an individual and as a member of a small group. Did you listen to what other groups had to say? What have you learnt about technology as a way to help people? As a class, discuss these points.

ICT Activity

After the publication of the report, the UNDP held a poll on their website to see if people thought that new technologies could make a difference to the poor.
The full report is at **www.undp.org/hdr2001/**
The poll is at **http://roo.undp.org/hdr/poll.cfm**
If you were to respond, how would you vote?
Use a PowerPoint presentation to explain why you would vote yes or no.

Is technology the answer?

Technology is part of the solution that can help global development. Some countries will be leaders in research and development. However, many more countries need to have access to these ideas to adopt and use the best technology. This means that poorer countries do not have to spend money on developing the technology but can get the benefits quickly.

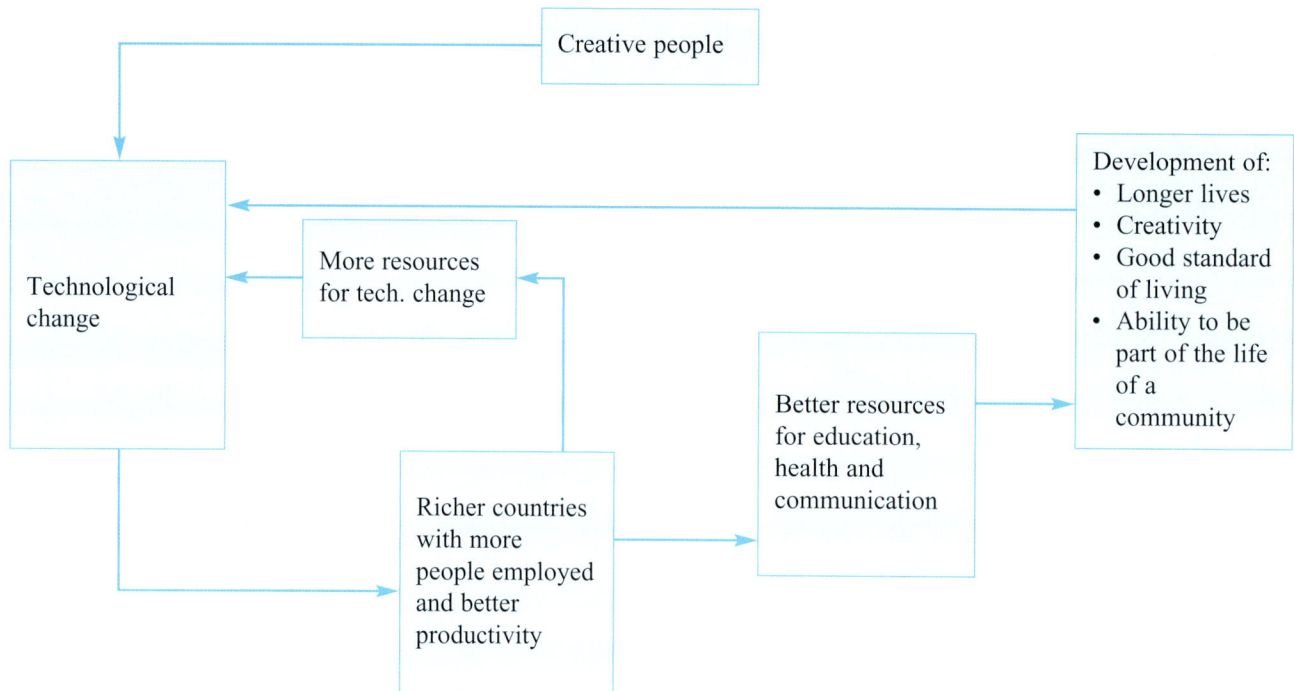

△ **Figure 2.16** Links between technology and human development

The cost of computers makes access to the World Wide Web too expensive for poor countries. In January 2001 the cheapest Pentium III computer was $700. This puts the Internet beyond the reach of poor and illiterate people. The Indian Institute of Science and Engineers at Bangalore developed a handheld Internet set for less than $200. It has touch screen functions and uses local languages to access the Internet. They hope to develop speech software for illiterate users.

△ **Figure 2.17** Breaking barriers to Internet access (adapted from *PC World 2000*; Simputer Trust 2000; Kirkman 2001)

◁ **Figure 2.18**
Solar power used to heat a kettle, Zanzibar, Tanzania

Activities

1 Study Figure 2.18.
a What is the man doing?
b How is he using technology?
c Why is he using this technology?
d How might this technology change his life?
e Do you think that this is a better way?
f Draw a simple sketch of the photo and use your answers to label it.

2 How would you heat water? What would life be like if you could not use this technology to heat water?

3 Read Figure 2.17. Use the flow chart below to show how Internet technology could help develop rural areas.

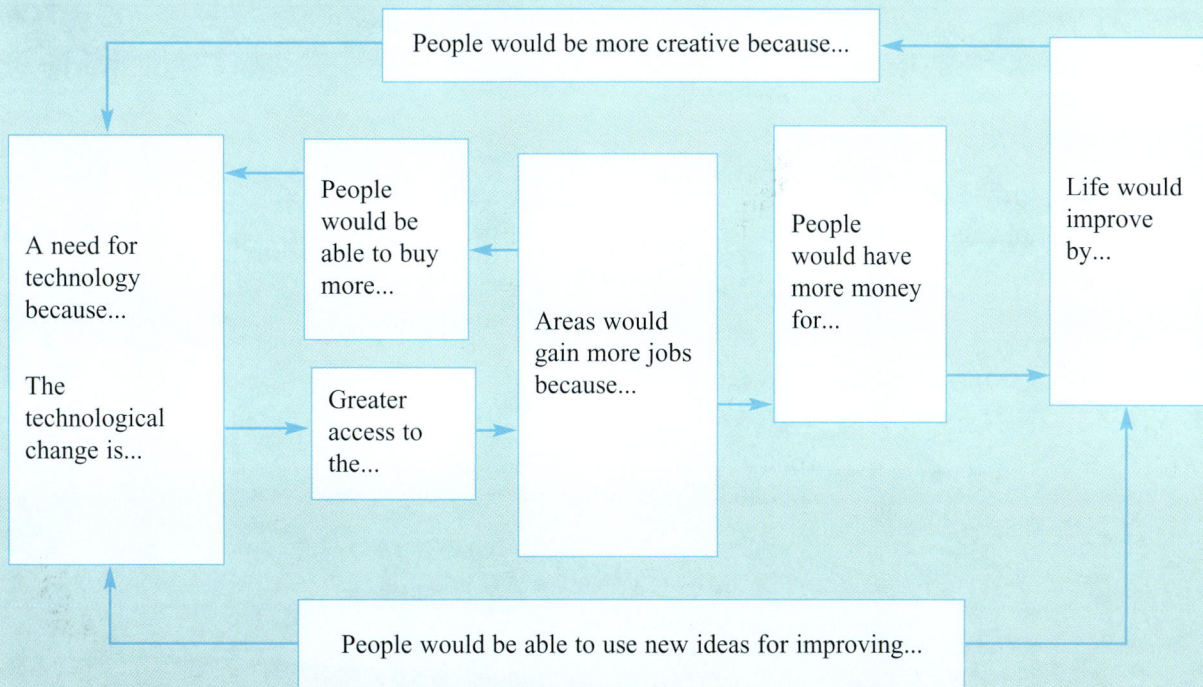

People would be more creative because...

A need for technology because...

The technological change is...

People would be able to buy more...

Greater access to the...

Areas would gain more jobs because...

People would have more money for...

Life would improve by...

People would be able to use new ideas for improving...

Most countries not on track to meet UN's 2015 goals

Human Development Report data indicates need for new initiatives

Mexico City, 10 July 2001 – Last September at the United Nations Millennium Summit, world leaders agreed on a set of goals for development and poverty eradication to achieve by the year 2015.[1] But, according to new analysis in the **Human Development Report**, many countries are not on track to achieve these goals.

- Eleven million children below age five still die every year from preventable causes – about 30,000 a day.

- Nearly one billion people still need access to safe drinking water.

- There are still 1.2 billion people who live on less than $1 a day.

[1] For the text of the Millennium Declaration, see www.un.org/millennium/declaration/ares552e.htm

Does aid have a role?

The use of the Internet in developing countries is increasing. However, a **digital divide** still exists between rich and poor countries. Internet costs are high in poorer countries. In the US, use of the Internet costs 1.2% of the average user's income. In Bangladesh, the same service costs 191% of a Bangladeshi user's income. Even an old invention like grid electricity is out of reach for 2 billion people.

At the 2000 G8 Summit protesters suggested that technology does not meet the needs of the poor. 'We can't eat computers,' complained the protesters. 'People are dying.' They burnt a laptop computer to make their point. Some fear that by emphasising technology, money will be taken away from other development projects.

Activities

1 Study Figure 2.19. Draw a poster to support the protesters' views about technology at the G8 Summit.

2 Study Figure 2.20. These numbers are so large that it is difficult to think about what they really mean.
a For each aspect, draw a diagram to compare rich and poor countries.
b Add a speech bubble to describe what you feel about this aspect.

Aspect	Poor countries (Developing)	Rich countries (Developed)
Education	845 million illiterate adults 325 million children not in school	50 million adults lack literacy skills
Income poverty	1.2 billion live on less than $1 per day	130 million have low incomes
Health	34 million have HIV/AIDS 968 million are without access to safe water 2.4 billion have no sanitation	1.5 million have HIV/AIDS
Nourishment	163 million children under five are underweight	8 million people are undernourished
Population	82% of the world's population 20% of the world's wealth	18% of the world's population 80% of the world's wealth

△ **Figure 2.20** Serious needs in many aspects of life

Assessment tasks

Can technology support development?

This chapter has looked at global development. To find out what this really means for people it helps to look at how technology has affected a country. India is a country that has experienced both positive and negative effects of technology.

India has a global technology hub at Bangalore (see Figure 2.4). India has the world's seventh largest number of scientists and engineers. India only ranks 63rd in the world technology achievement index. India's regional differences in using technology and a national adult illiteracy rate of 44% have slowed development.

India spends $2 billion per year on university education for 100 000 Indian professionals. A shortage of skilled workers in Europe, Japan and the USA has encouraged Indians to work abroad. Over 32 000 Indians moved to the USA between October 1999 and February 2000. Half of this percentage had computer skills. The loss of so many skilled workers makes it very difficult for India to develop the technology industries.

Assessment task – India and technology

The aim of this task is to describe what is helping India become technologically developed and what is stopping India becoming more developed.

1 Use an atlas to locate India and the UK on a blank world map. Use the atlas to compare the size of India with the UK.

2 From the work in this chapter, how developed is India compared to the UK? Give five examples.

3 List five positive effects of technology on India.

4 List three negative effects of technology on India.

5 Study the maps on page 42 (Figures 2.21–2.24).
a Draw a copy of the software map (Figure 2.21).
b Add the telecom backbone.
c Label and shade those areas with good access to technology **red**.
d Label and shade those areas with poor access to technology **blue**.
e Suggest a reason why areas have poor access.

Summary

6 Write a paragraph to explain how technology is helping India to develop. Use the 'I think …' frame below to help you. Use factual information to support your view.

I think that technology is helping/not helping India by . . .
Also I think . . .
Not everyone would agree with me because . . .
Also . . .
However, in conclusion I think . . .

△ **Figure 2.21** Software technology parks, India

△ **Figure 2.22** Telecom backbone network, Indi

△ **Figure 2.23** Per capita income of states (1997–8), India

△ **Figure 2.24** Literacy rate, India

Maps from www.mapsofindia.com/maps/india

Activities

1 Study the cartoon below (Figure 2.25). Use a dictionary to find out what the words mean.

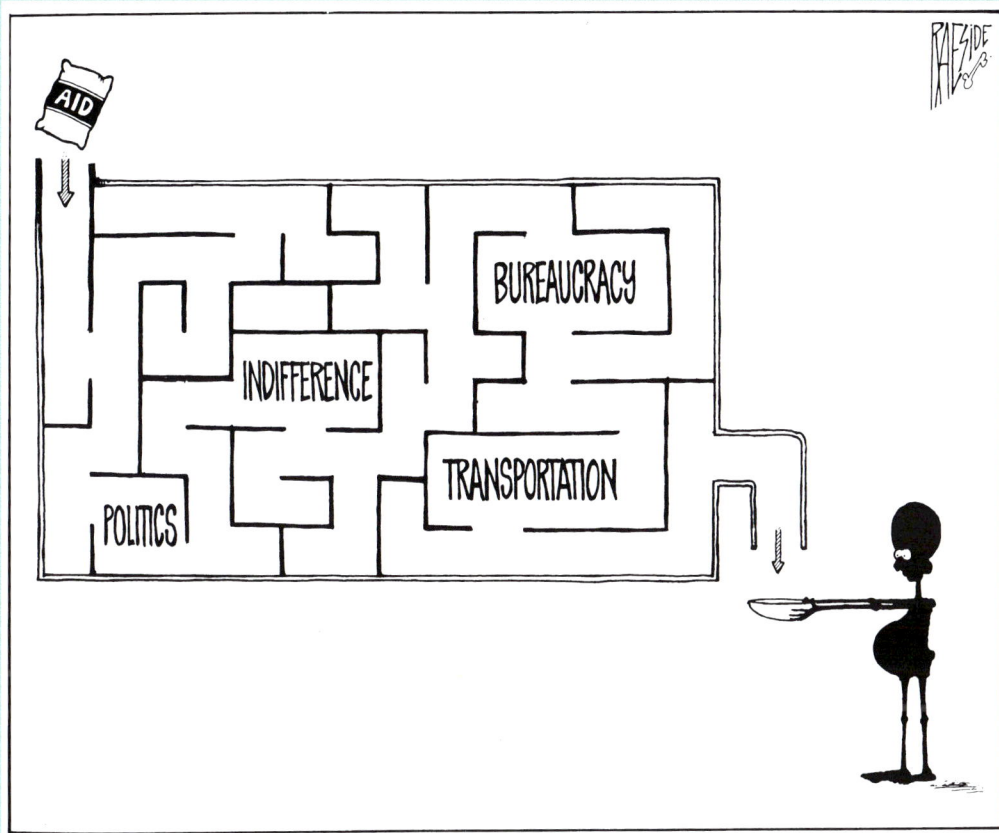

△ **Figure 2.25** Development is complex

2 What do you think the cartoon is saying about just giving aid to a country?

3 Redraw this cartoon. Add your own notes to explain what you think that the cartoonist is trying to say about development.

4 Draw a cartoon that shows technology as the most important way to achieve development.

5 Do you share knowledge or do you always keep ideas to yourself? Why is sharing knowledge and technology important if countries are to become more developed?

6 Look back at the experience map that you drew to represent your choices on page 25. Would you now change anything?

What is a natural hazard?

Natural hazards are natural processes or events that can cause harm to life or property. Earthquakes and volcanoes are called 'hazards' when they affect people. Both earthquakes and volcanoes are caused by movements within the Earth.

The study of these movements is called tectonics, so earthquakes and volcanoes are often described as '**tectonic activity**'.

Some countries have spent much money trying to stop natural hazards from affecting people.

▽ **Figure 3.1**

'I felt my heart beating harder and faster than ever before.'

'The sight of such appalling death and destruction was unbearable.'

'As the sky darkened, a strong wind blew toward us. We knew something terrible was about to happen.'

'I was shaking uncontrollably, it had affected me more than I thought.'

'All I could hear were the screams of those around me.'

The quotes and the photos in Figure 3.1 are linked to the following topics:

- **Living in a war zone**
- **Being near a volcanic eruption**
- **Driving in a Formula 1 race**
- **Riding a theme park roller coaster**
- **The effects of an earthquake**

The following words can also be linked to the photos and the quotes:

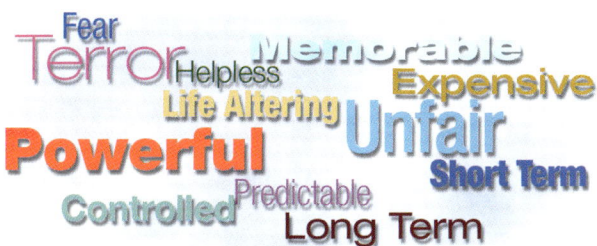

Fear
Terror Helpless Memorable
Life Altering Expensive
Powerful Unfair
Controlled Predictable Short Term
Long Term

Activities

1 Look at Figure 3.1 Match up each quote with the picture you think it fits best.

2 Use the words opposite. Choose the best descriptive words for each picture – you can use them more than once.

3 From the table below, choose the two best and the two worst things about living in a tectonically-active area.

▽ **Figure 3.2** Living in tectonically-active areas

	Benefits	Problems
Environmental	• Fertile soils made by volcanic eruptions • Tectonic activity creates amazing scenery	• Hazardous living conditions • Possible damage to ecosystem
Social	• All over the world communities have grown up next to natural hazards like volcanoes. These volcanoes form part of their religion and culture	• Damage to a community can take years to rebuild • Emotional damage to people who have suffered will never go or be mended
Economical	• Tourist attraction (e.g. Vesuvius and Pompeii) • Precious metals and gem stones are found in areas of tectonic activity	• Industry can be destroyed by tectonic activity • Crops can be lost and cattle killed

What do you already know about earthquakes and volcanoes?

The results of volcanic eruptions and earthquakes have shaped the world we live in. Even the hills and mountains of the UK were formed by tectonic activity millions of years ago. However, there are areas of the world that are being changed by dramatic volcanic eruptions and earthquakes today.

To help understand more about the lives of people in these areas, we must first find out how much we know about these types of event.

Facts and Opinions

- **Geography is made up of facts and opinions.**
- **People make decisions based on facts and opinions.**
- **Inaccurate facts and biased opinions can mislead decision makers.**

We can control the impacts of tectonic events to prevent loss of life.

In some countries a lot of money is spent each year to try and predict when earthquakes or volcanic eruptions are likely to happen.

As people have found out more about the structure of the Earth, the reasons for earthquakes and volcanoes have become clearer.

The environments created by tectonic activity can have special benefits for people.

In some parts of the world people live with the threat of these natural hazards from day to day.

People have been living in hazardous environments since the beginning of life on Earth.

Earthquakes and volcanoes create nothing but misery and suffering wherever they occur.

△ **Figure 3.3**

Geographers study the physical and human effects of volcanoes.

Geographers look at the following aspects of earthquakes and volcanoes: Where? How big? Why?

Human Factors

Living near volcanoes or earthquakes can affect how you live and work. Are you safe?

◁ **Figure 3.4** Ways in which volcanoes and earthquakes can be studied

Activities

1 Work in a pair. Write out the statements from Figure 3.3 onto strips of paper. Divide them into two groups – Facts and Opinions.

2 Choose one fact and one opinion and explain why you think it belongs to this group.

3 Why is it important to spot facts and opinions in things we read?

What causes tectonic activity?

Before we look at the effects of tectonic activity we need to investigate the causes. To do that we must start with the structure of the Earth.

Look at Figure 3.5. The Earth is a complicated structure. Most of the planet is a mixture of solid, semi-molten and liquid rock.

The surface of the Earth is called the crust. Figure 3.6 shows the crust in more detail. It 'floats' on top of the **mantle**. The crust is made from molten rock which has broken through to the surface and hardens as it cools.

Crust is still being made at places like the North Atlantic Ridge (see Figure 3.8). This is a **constructive plate margin**. Sometimes crust appears in one place called a 'hotspot'. Hawaii is a 'hotspot'.

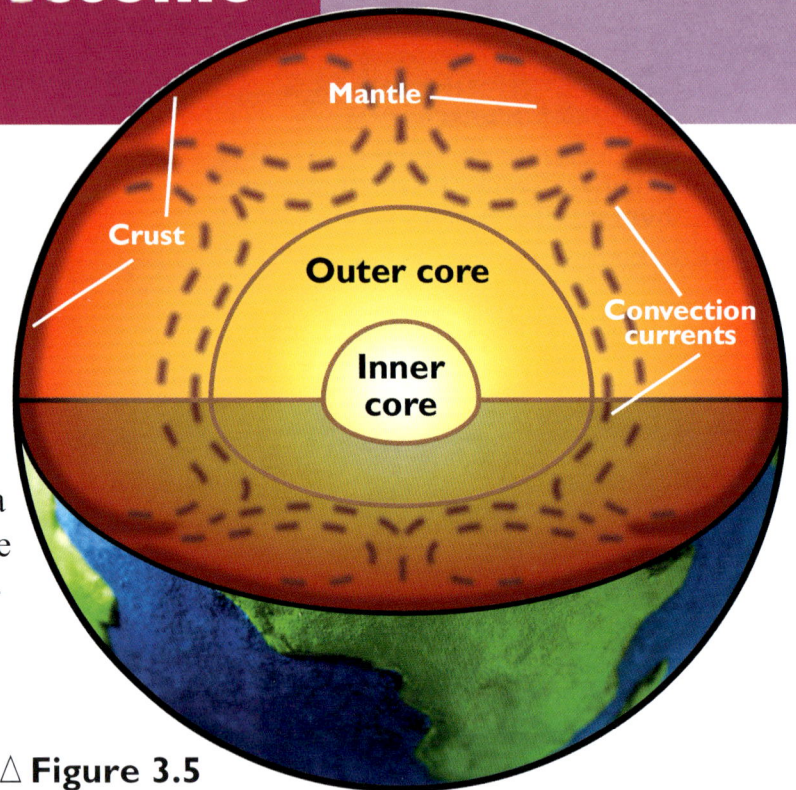

△ **Figure 3.5**
Inside the Earth

Some crust is taken back into the mantle. These areas are called **destructive plate margins** (Figure 3.7). Here, much tectonic activity takes place. The heat created can weaken the crust and lead to volcanic eruptions. Pressure can build up between the rocks, which break apart when the pressure gets too much. This causes earthquakes.

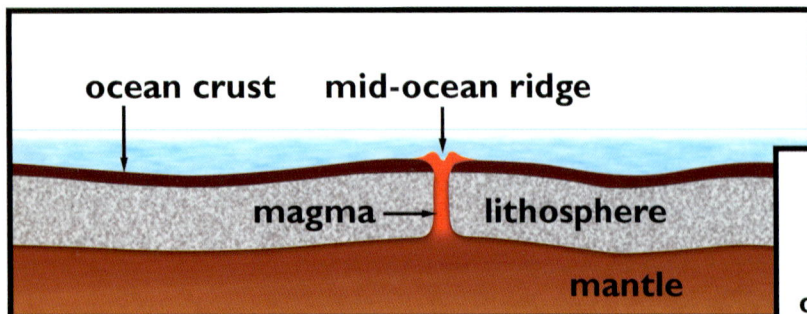

△ **Figure 3.6** Constructive plate margin or mid-ocean ridge

▷ **Figure 3.7** Destructive plate margin

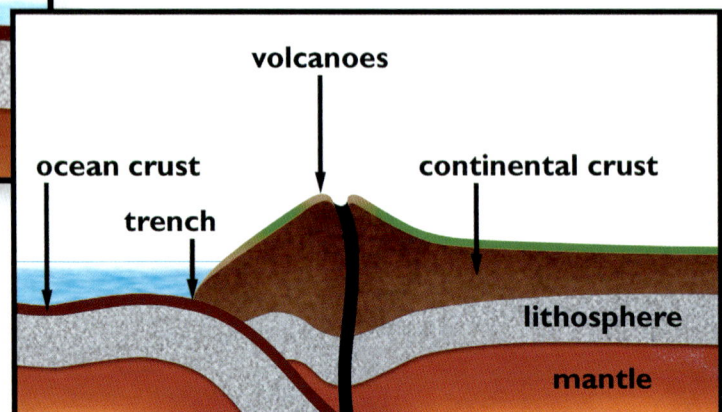

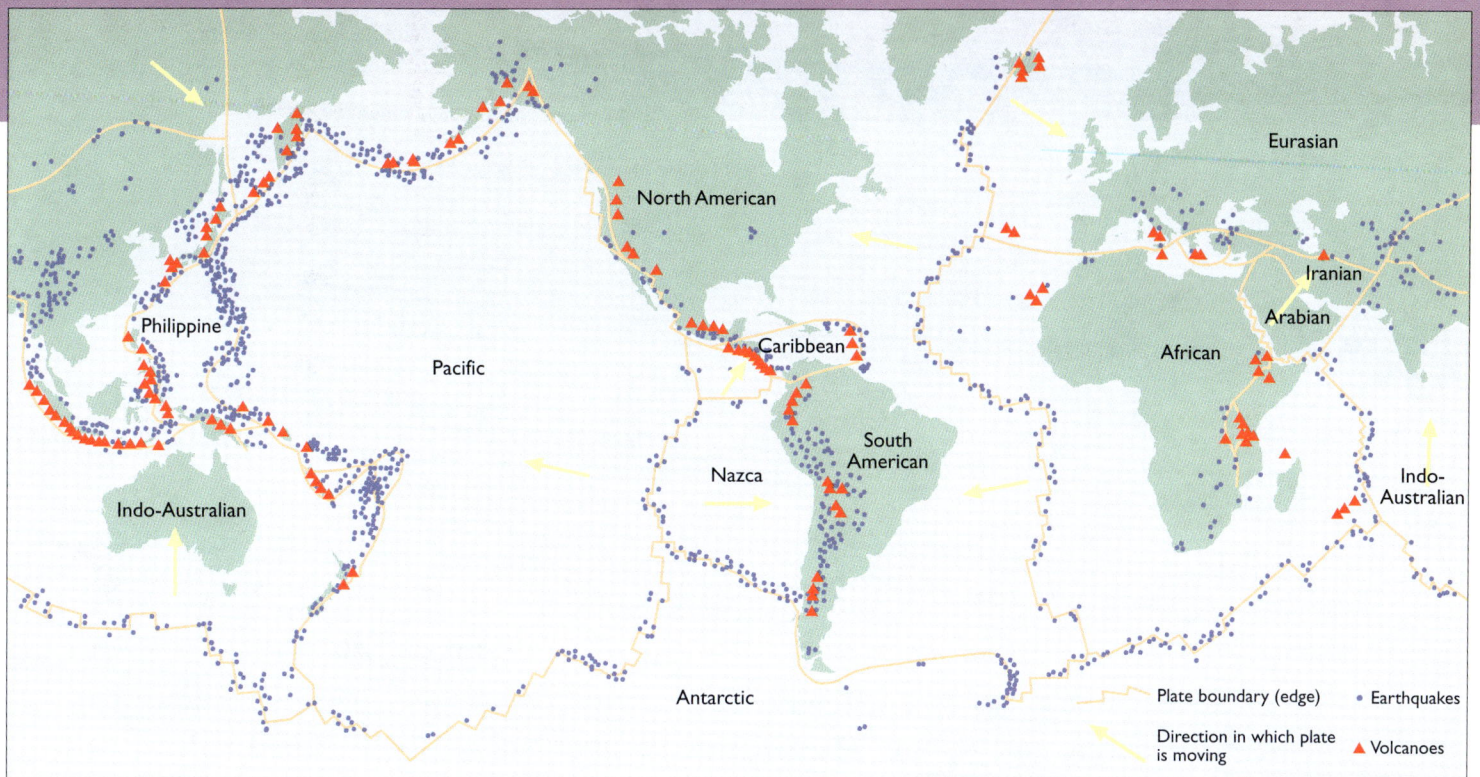

△ **Figure 3.8** Distribution of earthquakes, volcanoes and plate boundaries

The map (Figure 3.8) shows places in the world that are affected by either earthquake or volcanic activity. The pattern these places make matches with a map of the boundaries of **tectonic plates**. There is a close link between tectonic plates and tectonic activity like earthquakes and volcanoes.

The problem is, although we have a good idea of *where* the tectonic events might take place, it is still nearly impossible to know *when* things will happen.

ICT links

The following web addresses will help with your research into earthquakes and volcanoes.

www.pbs.org/wgbh/aso/tryit/tectonics/
http://volcanoes.usgs.gov/
http://earthquake.usgs.gov/

5W Activity

• **Hot spots**
• **Constructive plate boundaries**
• **Destructive plate boundaries**

WHAT are they?

WHERE do they occur?

WHY do they happen?

WHEN can they cause earthquakes and volcanoes?

WHO is most likely to see the results of these zones of activity?

Types of volcano

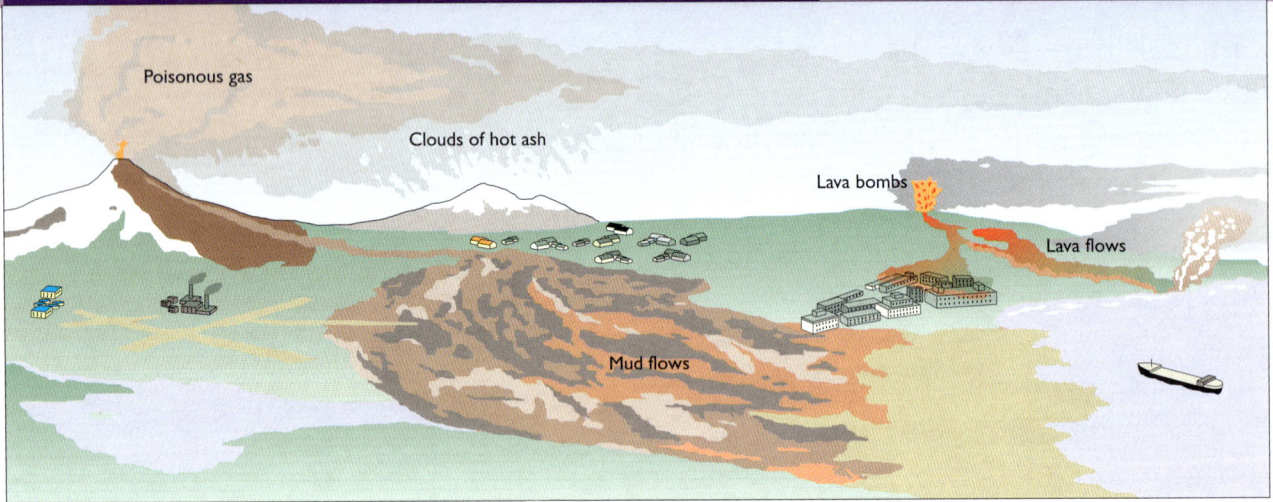

▷ **Figure 3.9** Problems caused by volcanic eruptions

Poisonous gas

Clouds of hot ash

Lava bombs

Lava flows

Mud flows

Volcanoes produce several different materials during an eruption. These include molten rock called **lava**, *lava bombs*, very *hot ash* and *poisonous gas clouds*. However, different volcanoes produce different materials.

At destructive plate margins, **strato volcanoes** are most usual. These are 'typical volcano shape'. At mid-ocean ridges and hotspots, **shield volcanoes** will form. These are much flatter and cover a wider area than strato volcanoes. The lava that erupts from these volcanoes is often more runny. This means it flows quicker and further than the lava of other volcanoes, covering wide areas.

△ **Figure 3.10a**

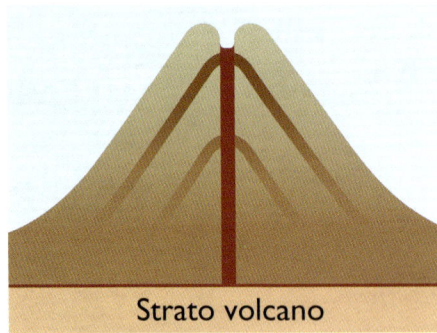

Strato volcano

△ **Figure 3.10b**

△ **Figure 3.11a**

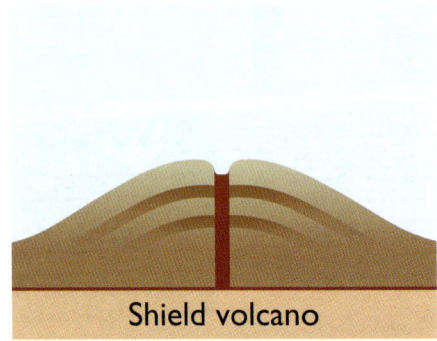

Shield volcano

△ **Figure 3.11b**

Activities

1 In a pair, draw a poster that shows the different types of volcanoes and the materials they produce.

2 In a pair, draw a poster that shows the dangers and problems of an erupting volcano.

About earthquakes

Earthquakes are vibrations caused by earth movements. Most happen because of plate movements (Figure 3.12). The vibrations can travel all round the world.

The place in the crust where the movement starts is called the **focus**. The place on the surface above the focus is called the **epicentre**. Vibrations travel outwards from the epicentre like ripples on a pond (Figure 3.13). Greatest damage is usually closest to the epicentre. The strength of the earthquake decreases away from the epicentre. Scientists can spot these shock waves from thousands of miles away.

❯ Measuring the size of earthquakes

There are two different ways of measuring earthquakes. The **Richter Scale** measures the energy released by an earthquake at its focus. The **Mercalli Scale** describes the damage caused by an earthquake (Figure 3.14).

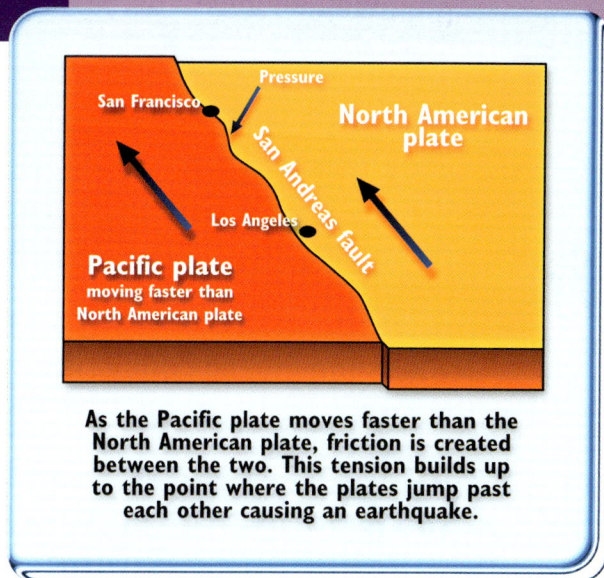

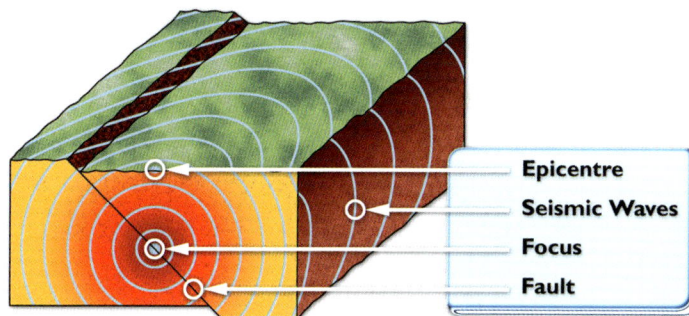

▽ Figure 3.12 A **fault** line

As the Pacific plate moves faster than the North American plate, friction is created between the two. This tension builds up to the point where the plates jump past each other causing an earthquake.

△ **Figure 3.13**

▽ **Figure 3.14** The Mercalli Scale

Less than 3.5	3.5 – 5.4	Under 6.0	6.1 – 6.9	7.0 – 7.9	8 or greater
Generally not felt, but recorded	Often felt but rarely causes damage	At most, slight damage to well-designed buildings. Can cause major damage to poorly-constructed buildings over small regions	Can be destructive in areas up to about 100 kilometres across where people live	Major earthquake. Can cause serious damage over large areas	Great earthquake. Can cause serious damage in areas several hundred kilometres across

Activities

3 Think about where you live. How would your area be affected by an earthquake?

ICT activity

Use the Internet to find out about earthquakes in England. Use www.bbc.co.uk/news to research the topic.

What is the impact of tectonic activity?

We are going to compare the effects of volcanoes and earthquakes in two different countries to see if there are similarities and differences in what happens after a tectonic event.

The examples are from Japan and Peru. As you complete the comparison you should ask yourself:

'Do volcanoes and earthquakes affect rich and poor countries in the same way?'

ICT links

Try to find out about tectonic activity in Japan and Peru.

You could use the following websites to get started, or if you have a CD-ROM encyclopaedia, try using that.

www.your-nation.com

www.census.gov

FACT FILE: PERU

Population 26.11 million people

GDP $11.20 billion

Area 1.29 million km^2

Birth rate 26.69 births/1 000 people

Life expectancy 69.97 years

Literacy 88.70%

Source: CIA World Factbook 1998

FACT FILE: JAPAN

Population 125.93 million people

GDP $3.08 trillion

Area 377 835 km^2

Birth rate 10.26 births/1 000 people

Life expectancy 80 years

Literacy 99%

Source: CIA World Factbook 1998

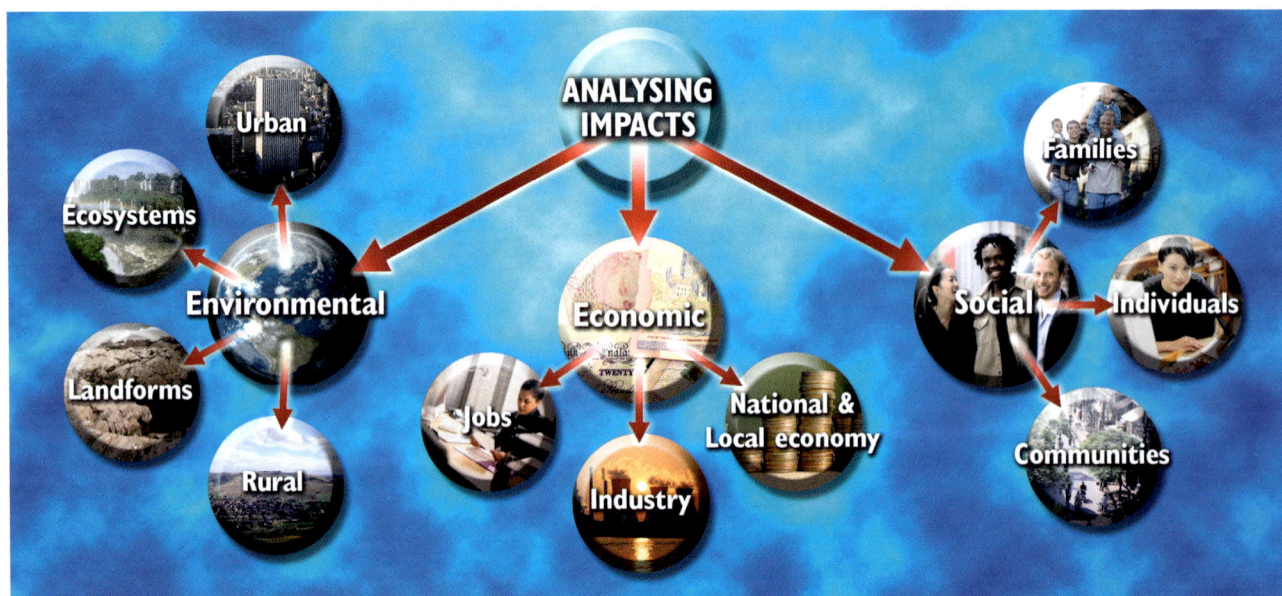

△ **Figure 3.15** Impacts of tectonic activity

To compare the impacts of earthquakes and volcanoes, try to classify the effects using the headings in the diagram above. This means you should put the effects into groups.

Usu Erupts

Friday 31 March, 2000 – Mount Usu volcano in Japan erupted today. A huge column of smoke and ash and a flood of volcanic ash and rock were spewed out toward a small nearby town.

Officials warned that the danger was far from over. 'Widespread damage could result from this eruption,' a spokesman said from the government's emergency headquarters in Tokyo.

About 51 000 people live in towns near Usu. Usu is 475 miles north of Tokyo on the island of Hokkaido. The eruption had been expected because of increased volcanic activity earlier this week. 11 000 people had been evacuated.

▷ **Figure 3.16**

△ **Figure 3.17**

Saturday 1 April, 2000 – Two more eruptions have rocked Mount Usu. The mountain is continuing to spew black smoke into the air. Scientists warn that another big eruption could be about to happen.

So far there have been no injuries, because people were evacuated in time. But some are worried that they might never return to their homes because they are gradually being buried in volcanic ash and rock. At least 18 000 local residents have been evacuated from the area. 3 000 soldiers, ships and helicopters are ready to help with more evacuations. Temporary shelters have been set up in schools and public halls.

Saturday 8 April, 2000 – Parts of the town are destroyed every day. Homes are covered in ash.

Landscapes look like the Moon, covered in ash. Roads have been twisted and cracked by the seismic activity. Buildings and cars are battered by falling rocks. Scientists said a new eruption might contain superhot lava, making it even more destructive than the earlier ones.

The central government is sending 1 500 temporary housing units for about 18 000 people camped out at evacuation centres.

– adapted from the BBC News Website

Earthquakes

Thousands left homeless in Peru!

At least 97 people were killed on Saturday 23 June 2001 when a huge earthquake struck Arequipa

The earthquake measured 8.1 on the Richter Scale. It was the worst to hit Peru for 30 years. At least 40 000 people in the south were affected. Many were made homeless. Most of the damage and casualties were around the city of Arequipa. Historic buildings were reduced to rubble. A tidal wave caused by the quake killed 39 people in Camaná. A powerful aftershock slowed the search for survivors.

From Ojo, Peru, 26 June 2001 – A great cloud of dust, walls destroyed, people walking about in the street, hungry and cold, this is the sad scene in this city. Villages have almost been wiped off the map. The people of Moquegua tried to retrieve any object or piece of furniture that might be useful. The town looked like it had been bombed.

From The Scotsman, 27 June 2001 – Forty-five minutes after Saturday's earthquake the first **tsunami**, 30 metres high, engulfed the mud huts of Camaná ... A second, bigger wave came moments afterwards. It ripped the fishing town to shreds ... Distressed villagers begged for help after their second night sleeping outdoors.

From Gestion, Peru, 26 June 2001 – Rebuilding in this part of the country will be much slower than in previous tragedies because there is less money available. The politicians must find some money to help.

From La Republica, Peru, 26 June 2001 – It is clear that what must happen now is to help those most affected. Many are living in dangerous buildings. Many lost everything. We must get aid to them straight away.

Adapted from *The Guardian* 30 June 2001

Many of the people who live in Usu and Arequipa know about the risks of earthquakes and volcanoes, but they still choose to stay. You would think that the problems caused by the earthquake and volcano would make people move away, but they haven't. Why is this?

The way people think about what might happen because of a natural disaster is called Hazard Perception. The way people respond to an earthquake or volcano depends on how they think it threatens them. If they think their lives will be put at risk again and again, they will probably move.

People may think the risks can be reduced by better emergency services, stronger buildings or evacuation plans. Although these things may help, there is nothing that can prevent natural disasters like volcanoes and earthquakes.

Less and More Economically Developed Countries (LEDCs and MEDCs) – Reminder!

Peru and Japan have different levels of economic development. Their income from exports helps with their ability to cope with the impacts of earthquakes and volcanoes.

So along with other things, a country's wealth is an important factor in dealing with these natural disasters.

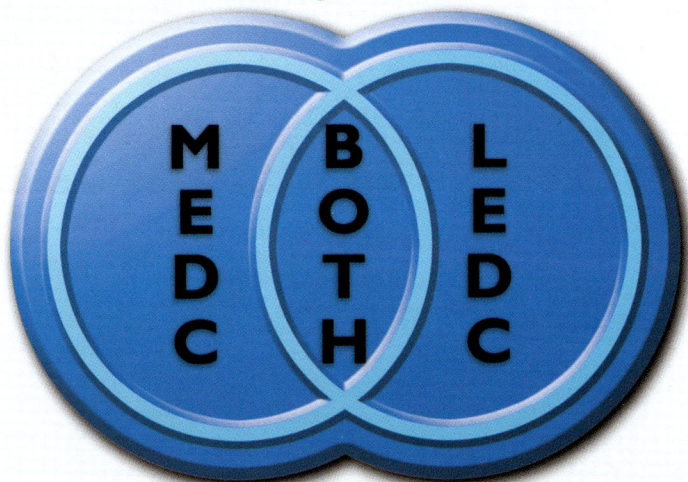

△ **Figure 3.18** Comparing the impacts of national disasters

Activities

1 Read with your teacher pages 52–4. Make two lists with the titles *The impact of tectonic activity on Japan* and *The impact of tectonic activity on Peru*.

2 Put your lists in order, with what you think are the greatest impacts at the top.

3 On a larger copy of the Venn diagram (Figure 3.18), sort out the impacts which only affect Japan (**MEDC**), those which only affect Peru (**LEDC**), and those which affect both.

4 Why are there differences between the lists?

What is disaster relief and how can it help?

For rich or poor countries, a natural disaster can cause serious problems for months and even years afterwards.

International relief organisations help to make life easier for people affected by disasters. These are often charities which use donations from individuals and companies. People affected by earthquakes will have different needs at different times. The things they need change from the time the disaster happens, to weeks and months after. Immediate, emergency relief is the responsibility of the individual country. For the people caught in the disaster, the most important needs are medical care, food, clean water and shelter.

Aid agencies can:
- provide emergency supplies
- offer advice on the best ways to search for any people who are still missing
- help to begin clean-up operations
- help educate people in basic first aid techniques for even longer term help
- give simple advice about creating stronger buildings.

However, it is up to the country to plan to avoid future problems.

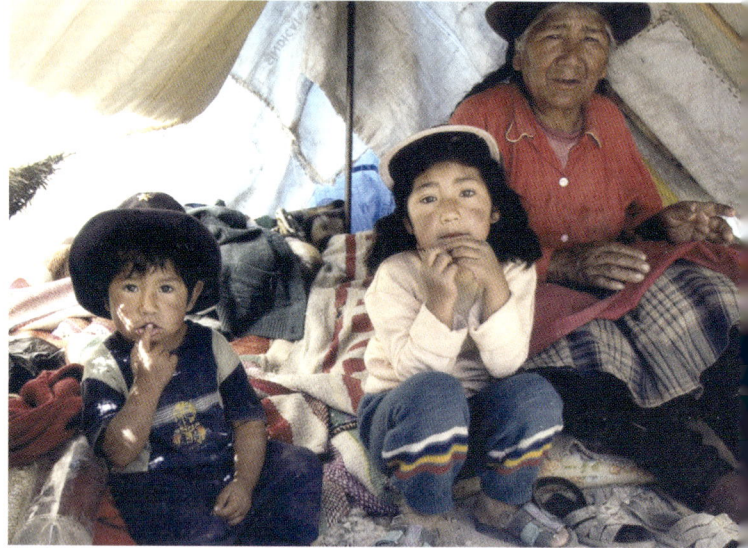

△ **Figure 3.19** People made homeless by an earthquake

▷ **Figure 3.20** Men inspecting rubble

A case study for disaster relief

Entire villages were flattened by the 7.9-magnitude quake in Gujarat state on 26 January 2001. Afterwards, hundreds of thousands of people were forced to sleep outdoors. There was little food and not enough clean water. Rebuilding has only just begun. Survivors still have the massive challenge of rebuilding their homes and shattered lives.

▷ **Figure 3.21**

The Christian charity, World Relief, quickly responded to the immediate crisis, providing funds for blankets, tents, food, water and cooking utensils. Then they started to think about long-term recovery and rebuilding but this could take years. As development work continues, World Relief plans to provide small business loans, called LifeLoans, to help poor families to earn a living.

△ **Figure 3.22** Wrecked buildings

Activities

1 Make a list of all the ways the people were helped on these pages.

2 Using the table below, sort your list into groups.

3 Which type of aid do you think LEDCs need most? Why?

Emergency Relief	Short-term Aid	Long-term Aid

How can we prevent volcanoes and earthquakes from causing so many problems?

Japan is an example of how people have tried to reduce the impacts of both volcanoes and earthquakes.

Japan is on the boundary between the Pacific and Eurasian plates (Figure 3.23). This means that earthquakes and volcanoes are always a threat.

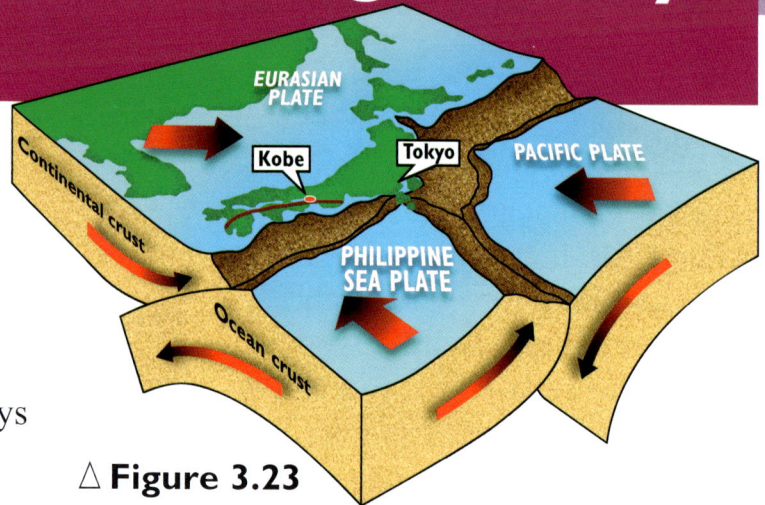

EURASIAN PLATE

Kobe

Tokyo

PACIFIC PLATE

Continental crust

PHILIPPINE SEA PLATE

Ocean crust

△ **Figure 3.23**

▷ **Figure 3.24**

Before	During	After
Prediction Research		
Hazard Education		
	Improved Buildings	
		Emergency Services

Figure 3.24 shows what the Japanese government has tried to do, before, during and after the event.

Japan has tried to predict when and where volcanoes and earthquakes are likely to happen. They can now warn people to get out in time.

Secondly, everybody in the country is taught about the problems caused by volcanoes and earthquakes. Japanese people even have a day's holiday to practise their emergency drills.

New buildings in the country have special and expensive foundations so they can move with the earthquake (Figure 3.25).

Finally, Japan has trained and equipped its emergency services. Police, Fire and Ambulance teams are able to work in the dangerous conditions found after an earthquake or volcano.

▽ **Figure 3.25** Foundations designed to move in an earthquake

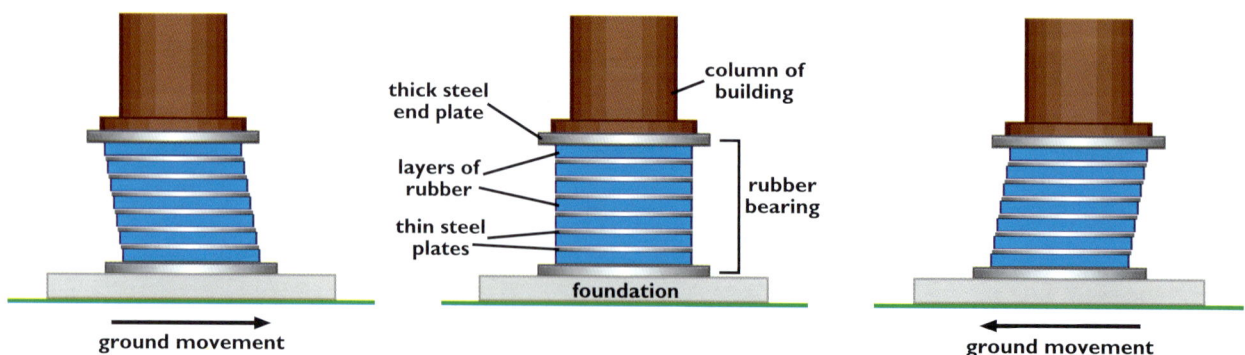

thick steel end plate

layers of rubber

thin steel plates

column of building

rubber bearing

foundation

ground movement

ground movement

Eyewitness Report

The earthquake that hit Kobe on 17 January 1995 caused severe damage to the city and the surrounding areas. Hundreds of buildings collapsed, roads cracked, bridges were destroyed, there were several landslips, and water, power and telephone lines were cut.

Even though the emergency services had been well trained, they found it very difficult to get to the worst hit areas. Without water, it was difficult to put fires out.

△ **Figure 3.26**

I had thought that all the news reports had prepared me for the damage but only when I was standing in the midst of 10 000 m² of ashes and twisted metal did the scale of the disaster really hit me.

Many people died and the cost of rebuilding is about £200 billion. Only the most recent buildings in the city had special foundations and even some of these 'earthquake-proof' buildings collapsed. We still can't beat nature.

No education could prepare people for such a devastating event. Even if they had predicted it, people would have been caught trying to escape and deaths could have been higher.

△ **Figure 3.27**

Activities

1 Read with your teacher the eyewitness report. Close your eyes and imagine the scene. Suggest one feeling you might have. Suggest one word to describe the scene.

2 Draw a sketch of Figure 3.26 in the middle of a blank page and **annotate** it with as much of the information as possible from the eyewitness report.

Philippine volcano spews truck-sized boulders

Tens of thousands of villagers have fled their homes as the Philippines' Mayon volcano unleashes a series of thunderous eruptions.

The volcano was spitting out flaming ash and huge great boulders. Witnesses said deafening booms rang out and giant cauliflower-shaped clouds of dust, ash and smoke shot up to 10 km into the sky, darkening this city of 120 000 people as well as surrounding towns.

About 23 000 villagers fled their homes as the series of explosions, which began on Saturday night, got worse on Sunday and shook villages up to 12 km away. 'The rocks coming down are as big as trucks,' **vulcanologist** Alex Baloloy said just before the first big blast at noon.

Rivers of fire – 'I heard what seemed like a huge thunder and I saw dark clouds ... then more boom-boom sounds,' a local journalist said. Moments later, fiery rocks and gas thundered down the volcano's slopes at 100 kph. A scientist said the 'rivers of fire' were 900° Celsius and would incinerate anything in their path.

△ **Figure 3.28**

Army trucks and police cars sped to try to evacuate the area. Some villagers fled on carts pulled by water buffaloes. In the panic, one woman had a heart attack and a pregnant mother gave birth. One villager, on a bicycle, was killed when a speeding van carrying rescue teams struck him. Another truck full of evacuees fell into a canal, injuring some of them.

△ **Figure 3.29**

Overcrowding – There is overcrowding and a lack of facilities in the evacuation centres set up in nearby towns. Up to 15 families have been forced to share a room in converted schools.

There are also reports of shortages of food and water at the centres. Emergency workers have been trying to stop people from heading back home. One relief worker said: 'No matter how much you try to stop them, they still want to go back.'

The precautions taken by the authorities meant that nobody was killed as a direct result of the eruption, in spite of its violence. This is a remarkable achievement in this area. The authorities are so confident of their ability to manage the effects of the eruption that they have even been trying to promote it as a tourist attraction.

Source: adapted from Erik de Castro, Reuters, Sunday 24 June, 01:26pm

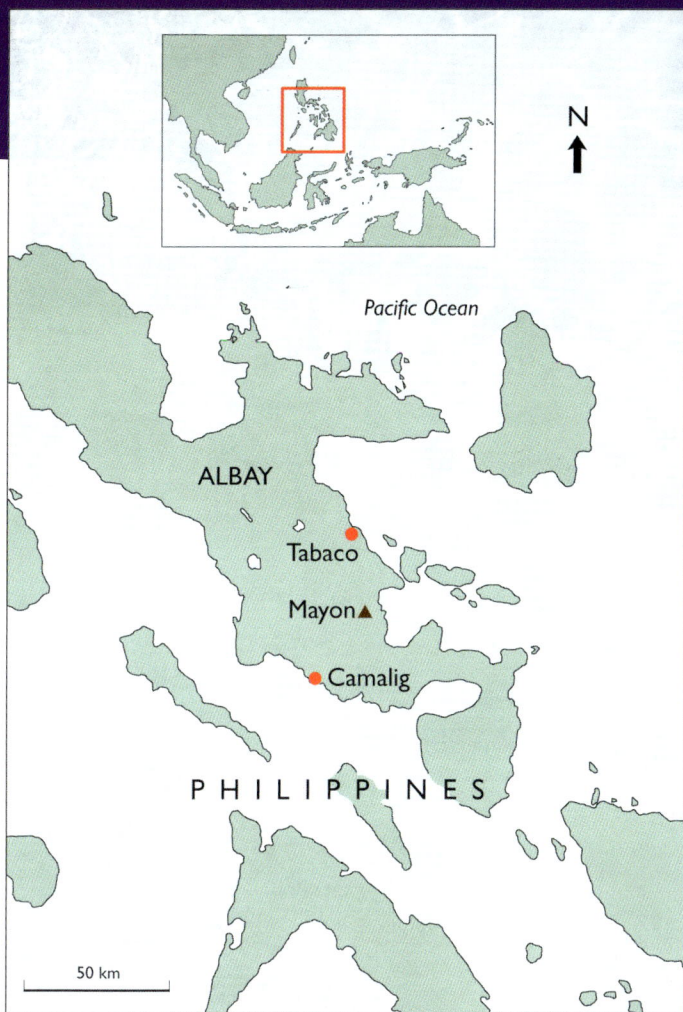

△ **Figure 3.30** Location map

▷ **Figure 3.31** Map of 1999 Mayon permanent danger areas

ICT link

Use this site to look at volcanoes in the Philippines

www.phivolcs.dost. gov.ph/Volcanoes/ Mayon/MayonIndex. html

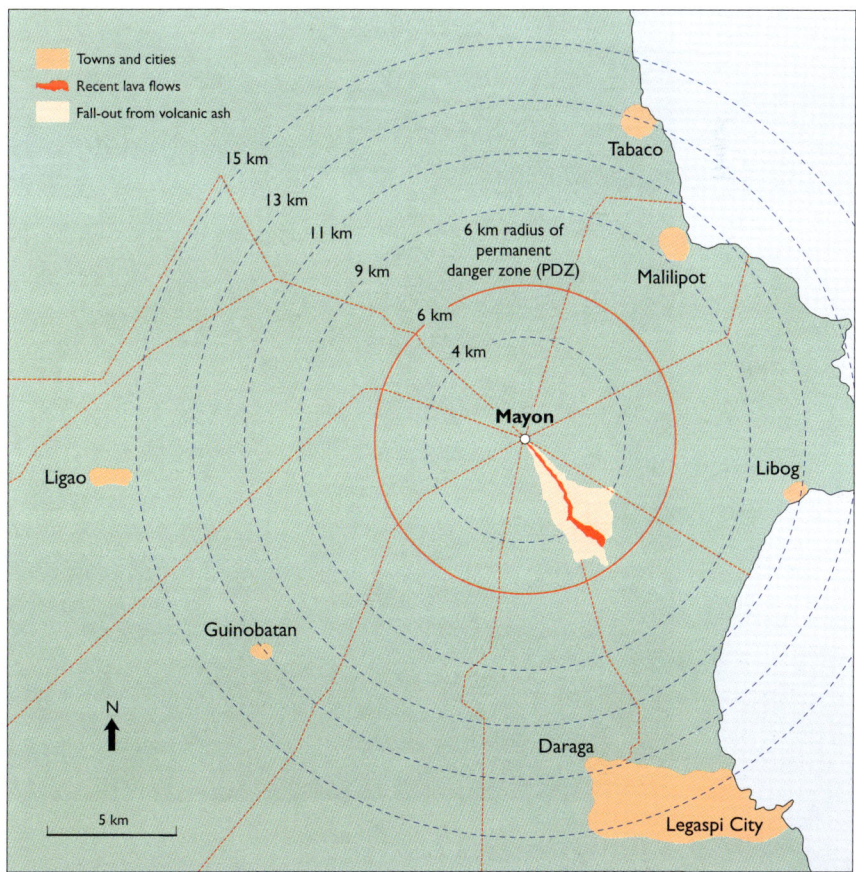

Assessment tasks

Tasks

Write a report to help the Philippines' government deal with the problems caused by the eruption of the Mayon volcano.

In a group, use the questions in Figure 3.32 to draw a concept map to get ideas for your report.

Now work on your own to write your report.

Use the following sub headings for your report:

- Where the eruption took place and why the Philippines will have more in the future
- What actually happened
- The effects on people
- The effects on property and business
- What could have been better about the way it was dealt with
- Suggestions for the future.

Make sure that you include diagrams, maps and pictures to illustrate your report.

Who?
...has suffered the worst as a result of the eruption?
...can provide aid?

What?
...are the three most important things to do next?
...aid do you need immediately?

Why?
...did the eruption happen?
...is it so difficult to help everyone?

Where?
...is the danger area?
...can people stay if they are homeless?
...is aid needed most?

How?
...has the eruption caused other types of hazard, like mudslides?
...can you keep people away from the danger area?

When?
...can you move people back to their homes safely?

△ **Figure 3.32**

Review

This review page is to help you reflect on your understanding of this topic.

The boxes show key words that you need to understand about this topic. Complete the Odd One Out exercise to check your understanding.

Copy each word and its number onto a separate card. Then put them into the order shown at the bottom of the page.

Decide which one is the odd one out from these groups and be ready to explain your reasons.

Key words

1 Tectonic plate	2 Volcano	3 Scientist
4 Epicentre	5 Focus	6 Mantle
7 Crust	8 Core	9 Fault
10 Earthquake	11 Hawaii	12 Hotspot
13 LEDC	14 MEDC	15 Disaster relief
16 Seismic activity	17 Richter scale	18 Magma
19 Lava	20 Constructive plate boundary	21 Destructive plate boundary

Groups

10	2	12
19	18	2
13	3	14
9	10	8
21	16	11
6	7	8

Where should we decide to live?

People all over the world will, at some time in their lives, have to make a decision about where to live. Here are some examples.

△ **Figure 4.1**

△ **Figure 4.2**

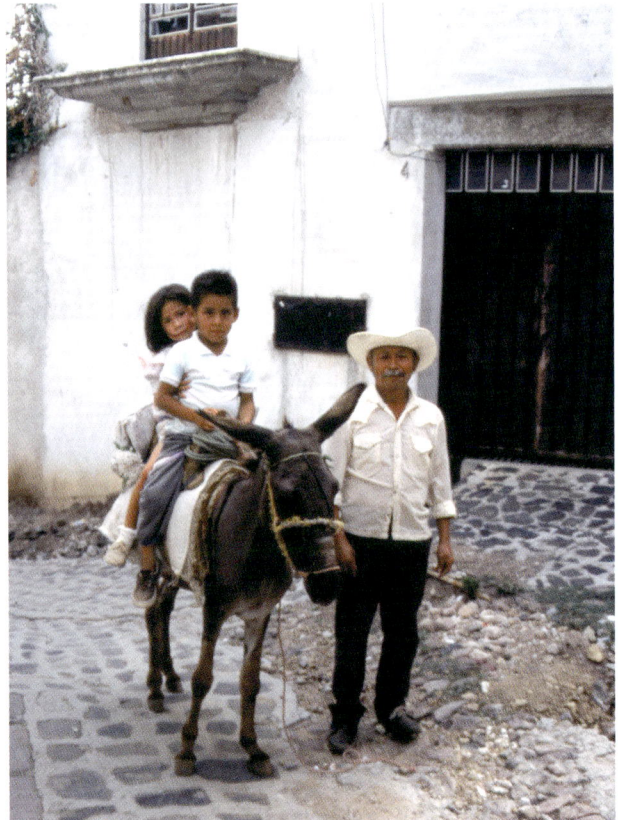

△ **Figure 4.3**

Activities

1 Look at Figures 4.1, 4.2 and 4.3 – but not the boxes below!

a List three things that might be similar about these people's lives.

b List three things likely to be different about their lives?

2 Now read the fact boxes below.

Dan and Amanda live in a cottage in the village of Hilderstone. They used to live in a nearby town. They were both born in large cities. They **migrated** to the village four years ago. They bought part of an old farmhouse and have restored it into a family home. Their two sons go to school in a nearby village, by car. They like the fresh air and quiet of the countryside. They both work in local towns. Many people in Britain are now moving out of towns to live in nearby villages. This is called **counter-urbanisation**.

These two men were born in a very poor village in West Africa. The photograph is over 100 years old. They were sailors. They had a dirty and hot job shovelling coal into ships' boilers. They worked on British steamships from Cardiff. They settled here because it was easier to get a job. There were people from their own country living in Cardiff already. They were **international migrants**.

Benito is the grandfather of José and Manuela. They have lived all their lives in the village of Tehuilotepec near the small town of Taxco. Taxco is 170 km south west of Mexico City. Their father has moved to Mexico City hoping for a better job. He will move his family there when he has made enough money. He is a **rural–urban migrant**.

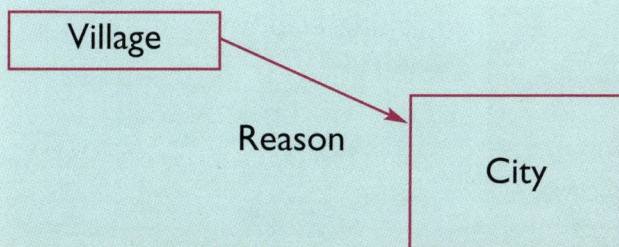

Village → Reason → City

a This diagram represents one of the fact boxes. Decide which one. Draw it and write the reason along the line.

b Draw a similar diagram with a reason for each of the other two fact boxes.

Is home sweet home?

- People have always needed to shelter by building a **settlement**. Settlements can be as small as a single house or as big as a large city.
- In some parts of the world many people live in single houses not close to others. These areas are often very **rural**. Most people live close to other houses, forming larger settlements.

- A small collection of rural houses is called a **hamlet**.
- A **village** is formed when there are more houses and a few services, such as a shop.
- Most people now live in **urban** areas in large towns or cities. Here there are many more services and **amenities**, such as a cinema. There are also better job opportunities.

△ **Figure 4.4** Desert village, Niger 15° 03' N, 5° 12' E

△ **Figure 4.5** Borth village, central Wales, 52° 59' N, 4° 3' W

△ **Figure 4.6** New York street scene, 40° 45' N, 74° 0' W

△ **Figure 4.7** African city, Ghardaia, 32° 31' N, 3° 37' E

▷ **Figure 4.8** Compass rose diagram

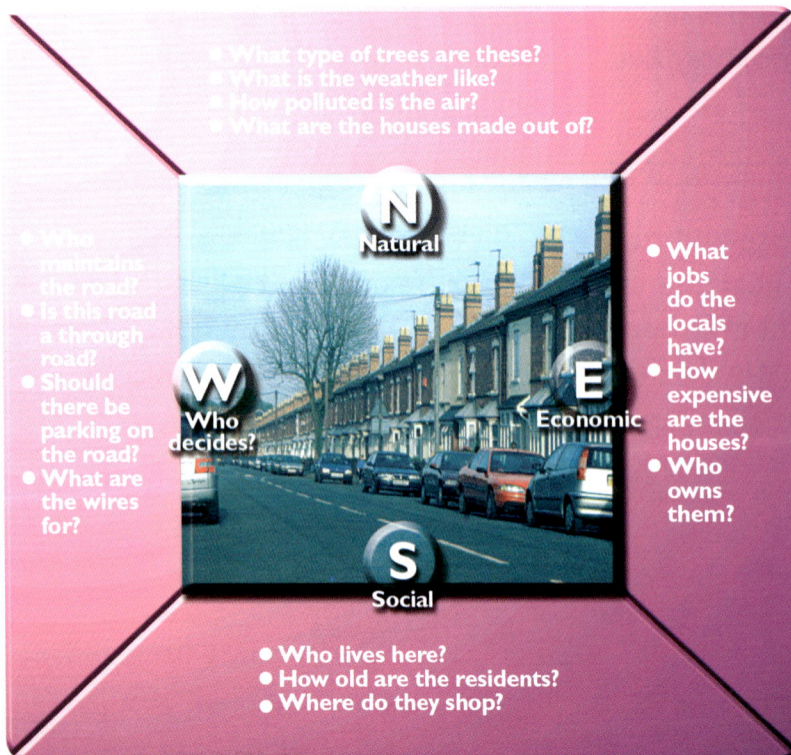

- What type of trees are these?
- What is the weather like?
- How polluted is the air?
- What are the houses made out of?

N Natural

- Who maintains the road?
- Is this road a through road?
- Should there be parking on the road?
- What are the wires for?

W Who decides?

E Economic

- What jobs do the locals have?
- How expensive are the houses?
- Who owns them?

S Social

- Who lives here?
- How old are the residents?
- Where do they shop?

Activities

Figures 4.4–4.7 show four settlements in different parts of the world.

a Which ones are rural and which are urban?

b Which ones are in MEDCs and which are in LEDCs?

2 Study Figures 4.8. This shows methods of looking at photographs in a different way. For each first letter of the four main compass points, write about what you see.

Will city life be for all?

Country	% population urban	growth rate % per year urban population	Workforce percentages		
			agriculture	industry	services
Australia	84.7	1.2	6	26	68
Burkina Faso	16.4	9.8	92	2	6
India	27.1	3	64	16	20
Japan	78.2	0.4	7	34	59
Mexico	73.6	2.5	28	24	48
Nigeria	40.5	5	43	7	50
UK	89.3	0.4	2	29	69
USA	76.3	1.2	3	26	61

△ **Figure 4.9** Selected statistics from countries at different levels of development

Activities

1 Study Figure 4.9.
a What does '% population urban' mean?
b Put the countries in order from the highest urban population to the lowest.
c Put the countries in order from the lowest workforce in agriculture to the highest.

d Compare the two lists. What do you notice about the two lists?

2 Which country's urban population has grown the most? Suggest a reason for this growth.

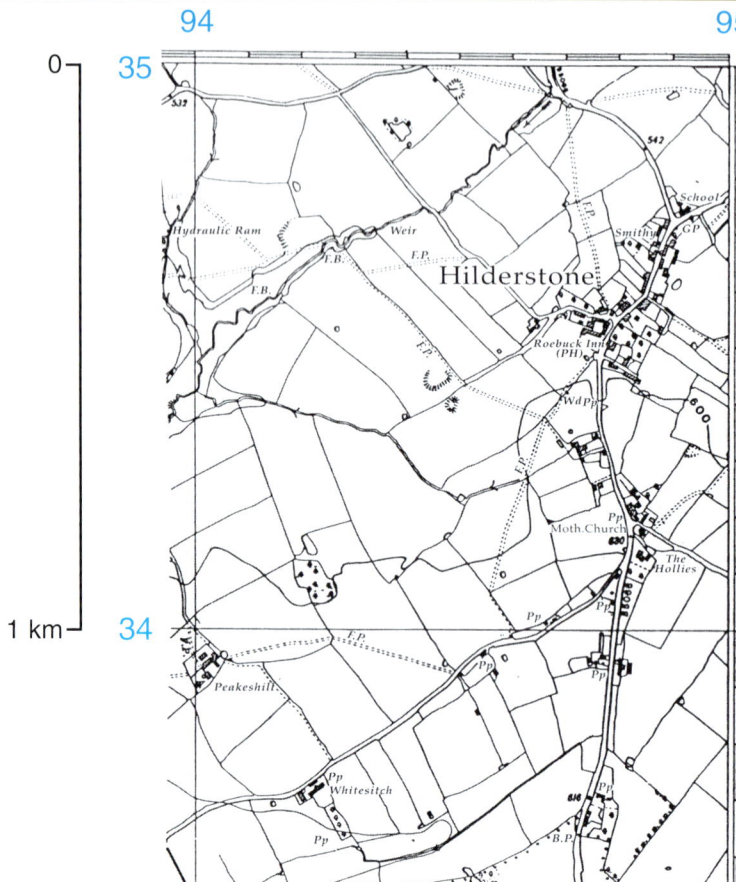

△ **Figure 4.10** Hilderstone 1949

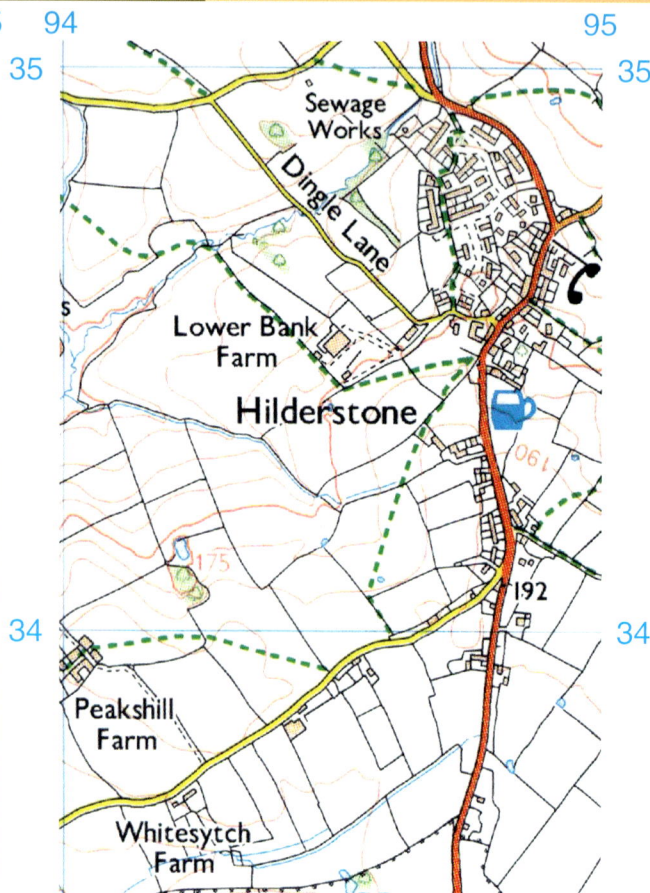

△ **Figure 4.11** Hilderstone today, from OS

© Crown copyright 2002

Hilderstone is a village with a population of over 400. It is located in central England, 11 km to the south east of the city of Stoke-on-Trent. There are farms in the area but most people do not have any links with farming.

Many people work in nearby towns. In the past most people either farmed or provided services for the local community.

Figure 4.12a shows part of a **directory** for the village for 1892. A hundred years later the 1991 **Census** (Figure 4.12b) shows a different picture of jobs done by the villagers.

COMMERCIAL.

Barker Frederick, shoe maker
Bloor John, farmer, Neville cottage
Bossen John farmer, Peak's hill
Bowers William, farmer
Bridgwood William, farmer
Cheadle Henry, farmer
Cope William, Horseshoe P.H. & carrier

Dalton George, tailor
Dalton Joseph, tailor
Fairbanks Alice (Mrs), shopkeeper, Sharpley heath
Fairbanks Wm. brick ma. Sharpley hth
Harthan Samuel, farmer, Stone heath
Heath Tomas, farmer, Stone heath
Heath William, sawyer, Stone heath
James George, wheelwright
Johnson William, farmer
Leedham Arthur William, farmer & landowner, The Lessows
Mayer George, farmer, Newfields & at Day hills
Meddings George, grocer, Post office
Meddins George, farmer, Green farm
Mountford John, farmer

Phillips Thomas, blacksmith
Porter Thomas, ownkeeper
Price William, farm baliff to John Bourne esq. jun. J.P.
Sharratt William, farmer, Stone heath
Shelley Thomas Holdcroft, Bird-in-hand P.H. Sharpley heath
Till George, bricklayer
Tunnicliff Lydia(Mrs.),butcher & farmer
Tunnicliff Thos. B. farmer, Whiessich
Turner Thomas, Roebuck inn
Udall George, farmer
Urion Daniel, farmer, Spot grange
Urion Frederick, farmer, Wooliscroft
Vernon John, farmer, Spot farm
Walkerdine Samson, farmer

△ **Figure 4.12a** Directory for village of Hilderstone, 1892

▽ **Figure 4.12b** Employment in Hilderstone, 1991

Employment	Agriculture	Energy & Water	Mining	Construction	Banking & Finance	Government & other services
Percentage	21%	7%	7%	27%	7%	31%

△ **Figure 4.14** Barn conversion

▽ **Figure 4.15** Bored children

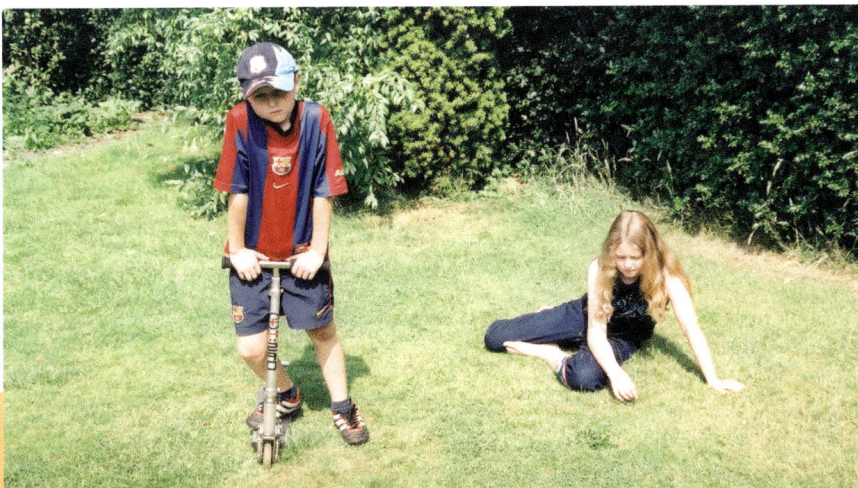

Activities

1 Study Figure 4.12a.
a Make a tally chart listing the jobs in Hilderstone in 1892.
b Put a cross next to the five jobs that you think are least likely still to exist in the village.
c Why might these jobs have disappeared?

2 Study Figure 4.10.
a Use tracing paper to copy the roads.
b Mark the buildings with blue dots.
c Put your tracing over Figure 4.11. Mark on the new buildings in red.
d List three differences between the two maps.

3 Put your tracing paper back over the map. Draw on a trail around the village. Try to stick to footpaths if possible and use arrows to point out changes that have taken place.

4 Young people in the village think that life here is boring! Are there any activities that teenagers can do here that cannot take place in towns and cities?

How good is public transport for villagers?

◁ **Figure 4.16** Bus timetable

8A
First PMT

HANLEY – LONGTON – BLYTHE BRIDGE – HILDERSTONE

HANLEY Old Hall Street (return via Lichfield Street, Birch Terrace to HANLEY Bus Station), Lichfield Street, Victoria Road, Fenton, King Street, LONGTON Market Street (return via The Strand), Uttoxeter Road, Meir, Uttoxeter Road, Catchems Corner, Lysander Road, Farnborough Drive, Grindley Lane, Blythe Bridge, Draycott, Cresswell, The Hunter, Saverley Green, Fulford, Townend, Cross Gate, Moss Lane, Mossgate, Bird-in-Hand, HILDERSTONE.

Monday to Saturday

		NS	NS	S	NS		
HANLEY, Argos Superstore	....	0841	1041	1241	1341	1541	1741
LONGTON, Market Street (Stop F)	....	0856	1056	1256	1356	1556	1756
MEIR, Broadway	....	0902	1102	1302	1402	1602	1802
BLYTHE BRIDGE, Rail Station ☏	0732	0907	1107	1307	1407	1607	1807
Draycott, Church	0735	0910	1110	1310	1410	1610	1810
Cresswell, Rookery Crescent	0738	0913	1113	1313	1413	1613	1812
Saverley Green, Greyhound	0740	0915	1115	1315	1415	1615	1815
Fulford, Village Hall	0743	0918	1118	1318	1418	1618	1818
Mossgate, Moss Lane	0745	0920	1120	1320	1420	1620	1820
HILDERSTONE, Old School	0750	0925	1125	1325	1425	1625A	1825

	S	NS	NS	S	NS	S	NS	
HILDERSTONE, Old School	0749	0754	0934	1034B	1134	1334	1434	1634B
Mossgate, Moss Lane	0753	0758	0938	1038	1138	1338	1438	1638
Fulford, Village Hall	0756	0801	0941	1041	1141	1341	1441	1641
Saverley Green, Greyhound	0759	0804	0944	1044	1144	1344	1444	1644
Cresswell, Rookery Crescent	0801	0806	0946	1046	1146	1346	1446	1646
Draycott, Church	0803	0808	0948	1048	1148	1348	1448	1648
BLYTHE BRIDGE, Rail Station ☏	0807	0812	0952	1052	1152	1352	1452	1652
MEIR, Broadway	0813	0818	0958	1058	1158	1358	1458	1658
LONGTON, The Strand (Stop B)	0820	0825	1005	1105	1205	1405	1505	1705
HANLEY, Bus Station	0835	0845	1025	1130	1225	1425	1525	1725

Notes:
A – Saturday only continues to Milwich
B – Saturday only commences from Milwich at 1630
NS – Not Saturday
S – Saturday only

Most journeys extend to/from Chell via High Lane

Sunday and Bank Holidays: No service

20.9.99

Page 12

▽ **Figure 4.17** Bus map

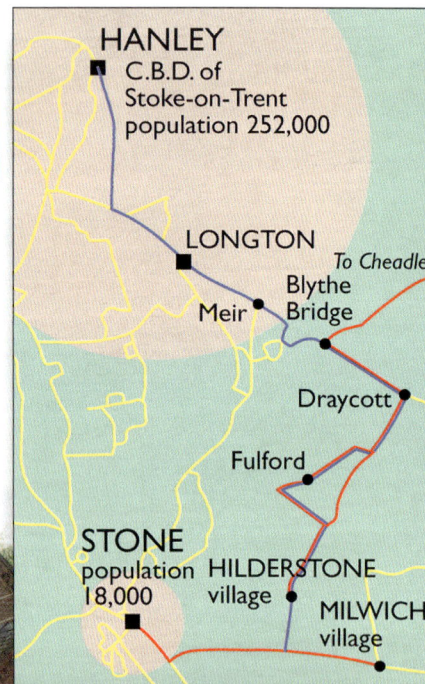

N

Key:
- Built-up area
- Route 8A
- Route 249
- Other routes
- ■ Main towns
- • Villages

0 1 2 3 4 5 miles
0 1 2 3 4 5 6 7 8 km

HANLEY
C.B.D. of Stoke-on-Trent
population 252,000

LONGTON
Meir
Blythe Bridge
To Cheadle
Draycott
Fulford
STONE
population 18,000
HILDERSTONE village
MILWICH village

△ **Figure 4.18** A long wait?

▷ **Figure 4.19** The village bus arrives at last!

Hilderstone is a village with a population of over 400. It is located in central England, 11 km to the south east of the city of Stoke-on-Trent. There are farms in the area but most people do not have any links with farming.

Many people work in nearby towns. In the past most people either farmed or provided services for the local community.

Figure 4.12a shows part of a **directory** for the village for 1892. A hundred years later the 1991 **Census** (Figure 4.12b) shows a different picture of jobs done by the villagers.

Activities

Refer to Figure 4.16 in this activity.

1 How many places does the bus call at?

2 What time is the earliest bus you can catch to Hanley from Hilderstone on a weekday?

3 What time is the last bus you can catch from Hanley on that day?

4 Why do you think that the bus only runs Monday to Saturday?

5 Do you think it is a good idea to use small buses for routes like this one?

6 Which groups of people rely on local buses?

◁ **Figure 4.20** Teleworker

What changes are taking place in the city?

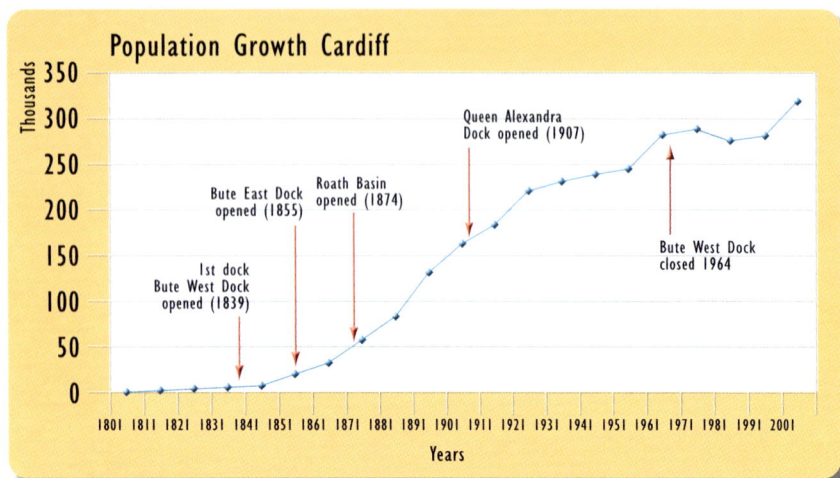

△ **Figure 4.21** Population growth graph, Cardiff

▷ **Figure 4.22** Tiger Bay, Cardiff

△ **Figure 4.23** Bute Street, Butetown, before re-development

The village of Hilderstone has two bus services to the nearby town of Stone and the city of Stoke-on-Trent. Figure 4.16 shows a timetable for one. Stone

A city of immigrants

(population 18 000) has two supermarkets, banks and a small range of clothes and shoe shops. Stoke-on-Trent (population 252 000) has its **Central Business District (CBD)** at Hanley. Here there is a large covered shopping centre with all the usual shops found in our big cities. Hanley also has banks, cinemas and theatres.

Decline of the docks

There is a rail station with hourly links to London.

Hilderstone has a population of about 400. There is no shop or post office. 11% of households do not own a car but half have more than one car. Most people travel to work by car. Some people **car pool**, which means sharing a car journey to work. 13% of the people in this village and the surrounding area work at home. Many **telework** using telephones, computers and email.

△ **Figure 4.24** Many local residents worship in the mosque. This is an old mosque, one of Britain's first. This photo is from the 1970s

Here is an example of how one area of a city has changed quickly in recent years. The city is Cardiff (population of over 300 000), in south Wales.

Cardiff docks were built to export coal. Many people migrated here in the 19th century to find jobs. Many of the **immigrants** came from villages, looking for a better life. Cardiff also attracted African sailors. They settled in an area close to the docks called Butetown. This area was also nicknamed 'Tiger Bay'.

Cardiff docks stopped exporting coal in the 1960s. The **residential** area of Butetown became run-down. The older houses were unfit to live in. Butetown was re-developed.
• Old housing was cleared
• Modern high-rise blocks of flats were built
• A new park was created

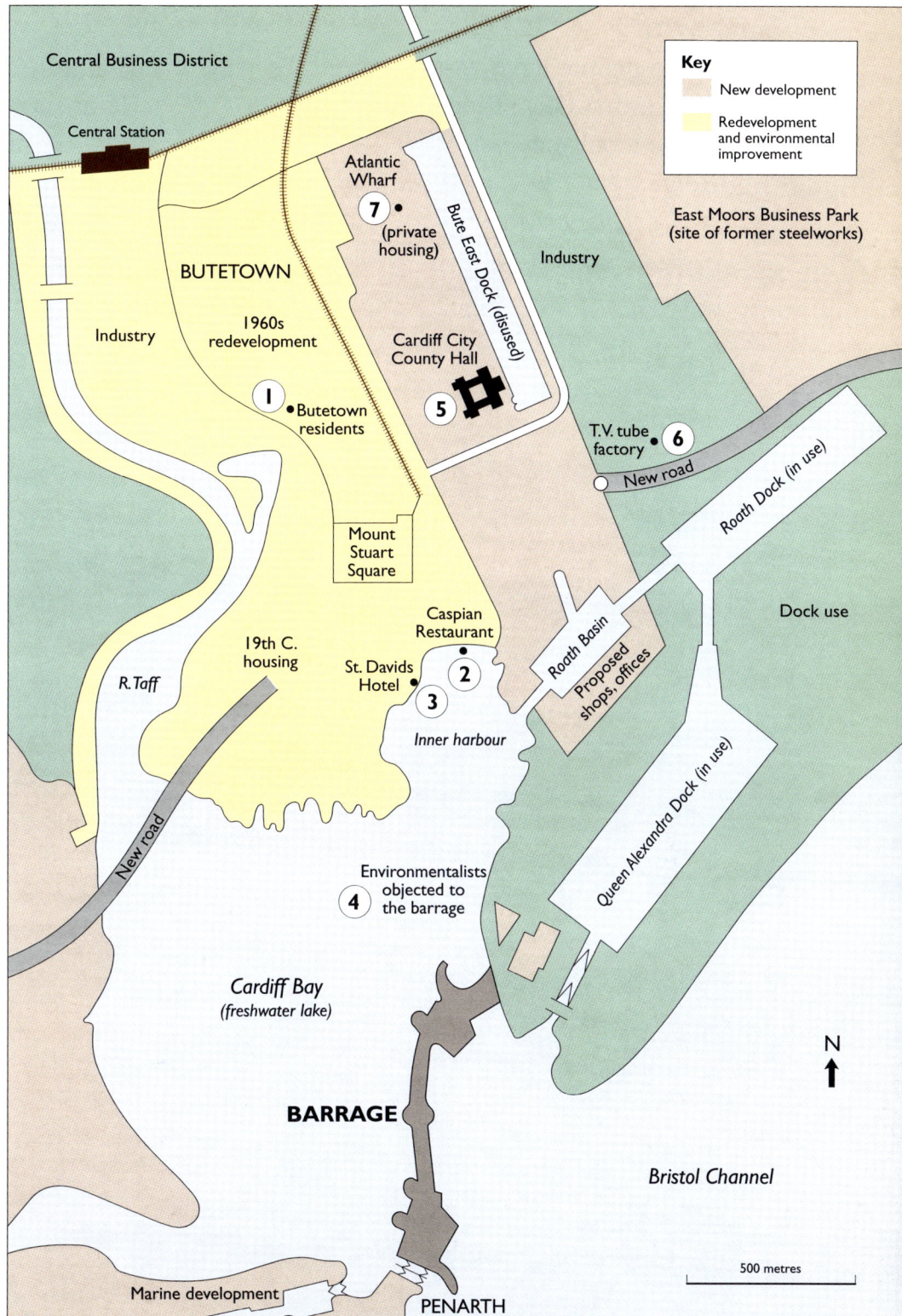

△ **Figure 4.25** Map of Cardiff Bay

What changes are taking place in the city?

Has the re-development been a success?

Some people have worries about the way the Cardiff Bay scheme has developed. Some locals have said, '*It's not for us ... we get few jobs ... we can't afford to go to the pricey restaurants*'. Others have felt that the development is the best thing to happen to the area '*before this we had mudflats, **derelict** buildings and a poor image ... it's put the area on the map. No more Tiger Bay but the new Cardiff Bay!*'

Activities

1 Read together about these seven groups interested in the Cardiff Bay area.

Manager of St David's Hotel

This is Wales' first five-star hotel with a health spa. The location is great.

Butetown residents

We already lived here before the scheme. Male unemployment of 50% is one of the highest in Wales.

Environmentalists

We never wanted the barrage in the first place. Seabirds do not come now the mudflats have gone.

Owner of the new Caspian restaurant

I have rented a brand new building built over water with good views.

Atlantic Wharf residents

We moved here because we wanted to be near the city centre. The houses and flats are new but we will probably move on in a few years time. We do not know any local people as they live the other side of the railway line.

Cardiff City Council

We now run the planning of this area. Our Council Offices are here. Having the new Welsh Assembly here is also a great boost for the area. Cardiff will benefit from more jobs and more money for the City Council from council tax.

Manager of Schott/NEG TV tubes

We came here because of the excellent site close to many other hi-tech firms in south Wales. There is also a readily available skilled workforce.

▷ **Figure 4.26** The flats built in the 1960s

△ **Figure 4.27** Atlantic Wharf – new private housing

△ **Figure 4.28** New restaurant built on the Bay

▽ **Figure 4.30** The barrage turns the Bay into a freshwater lake

△ **Figure 4.29** St David's Health Spa and Hotel

Activities

2 **a** Which groups support the changes to Cardiff Bay?
b Which groups are against the changes to Cardiff Bay?

3 **a** Choose one group.
b List two reasons why they would like the development.
c List two reasons why they would not like the development.
d Suggest one thing they would change. Suggest why.

4 Do you think the development has been good or bad for Cardiff? Suggest why.

How do Mexicans live in town and country?

Mexico is a large country in North America. The population is around 93 million. The country became independent from Spain in 1821. The official language is Spanish, although Native American languages are spoken in some rural areas. The climate varies from a dry desert in the north to rainy tropical forest in the south east. The border with the USA is over 3 000 km long. Oil and tourism earn Mexico the most foreign income. Mexico is a developing country with a yearly income of about $4 000 per head.

The Zócalo is the central square of Mexico City (Figure 4.31). The Native American dancers can trace their heritage back to the time of the Aztecs over 600 years ago. The buildings show the influence of the Spanish. The Spaniards conquered the Aztecs in 1519 and destroyed their large capital city of Tenochtitlan close to the Zócalo.

Most people in Mexico now live in urban areas (74%). The population of Mexico is rising because the annual **birth rate** is higher than the **death rate**. Mexico's birth rate is 25 per thousand of the population per year and the death rate is only 5. The birth rate is higher in rural areas. Children help in the fields and with housework. The death rate has fallen because now there is better health. 93% of Mexicans have access to health services.

△ **Figure 4.32** Ruins of Tenochtitlan, the Aztec capital

▽ **Figure 4.31** Indian dancers in the Zócalo

Activities

1 With a partner write down three things you already knew about Mexico.

2 Now write down three things you have found out about Mexico from this page.

Tehuilotepec is a village 5 km north of Taxco. Many people have left the village to live and work in larger towns or cities. **PUSH** and **PULL factors** explain why people leave villages. People are often pushed from a village because there are not enough jobs and living conditions are poor. They are pulled to the towns and cities because they can get better jobs and they hope they might find a better lifestyle for their families. The pull is often about what people *think* will be better. They may have seen a glamorous TV programme of life in the city or heard tales from a visitor who has moved there.

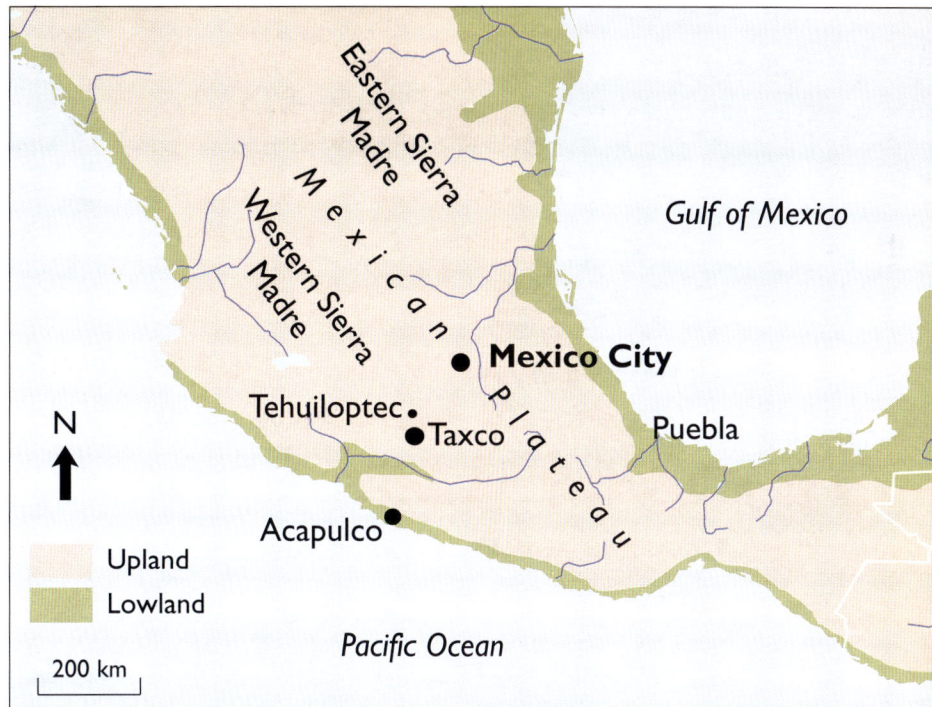

△ **Figure 4.33** Location map for Mexico

Activities

3 Here are some reasons why people might leave their village. Copy each reason onto a separate piece of paper. Sort the reasons into PUSH and PULL factors.
- There's not enough food to go around.
- You can easily get an **informal job** such as shoe-shining.
- There is a chance to get a steady job in a factory.
- There is only a primary school.
- There are only a few low-paid jobs.
- Hospitals are nearby.
- Health care is basic.
- There is always something to do.
- There are good educational opportunities.
- It's not very exciting for young people.

△ **Figure 4.34** Shining shoes is an informal job that any new migrant can take up

Village life versus the city?

△ Figure 4.35 'The Village Scene', a print sold to the tourists

△ Figure 4.36 The village today

Until recently many people born in the villages of Mexico lived there all their lives. Today there are more opportunities for young people. The activities below show some of the choices now available.

Activities

1 You are a young person from Tehuilotepec. You have just left school. Should you move or stay? Here are some options:

Stay in the village (Figure 4.36). Your parents need you to help on their small farm. You can travel every day to a job in the nearby town of Taxco. Your farm could provide more food for the hotels and restaurants in Taxco.

Move to Taxco (Figure 4.39). This town is an old silver mining centre with over 300 silver shops. Tourists now arrive by coach from Mexico City. You could sell tourist souvenirs like the print (Figure 4.35). These are made by your family in the village.

Move to Mexico City (Figures 4.38, 4.40 and 4.41). This city is one of the largest in the world with over 19 million people. It's a three-hour bus journey away. There are plenty of opportunities but you may only get an informal job such as shoe-shining. Housing is difficult to find. You have an uncle who lives in a large **shanty town** on the edge of the city. The city has an enormous traffic problem that causes air pollution. There are also problems with water supply and sewage.

a Suggest two problems if you stayed in the village. Draw and describe the village (Figure 4.36).

b State two advantages to working in Taxco. Draw and describe the village of Taxco (Figure 4.39).

c State two advantages to working in Mexico City.

d Suggest two problems if you went to Mexico City.

2 If you had the choice, where would you go? Write a brief letter to your family explaining the reasons for your choice.

△ **Figure 4.37** Mexico City is over 2 200 metres above sea level 2 000 km². It covers an area of over 2 000 km² and houses more than a fifth of Mexico's entire population

△ **Figure 4.38** Mexico City shanty towns using the steep unwanted land

▽ **Figure 4.40** The Latin American Tower

△ **Figure 4.39** Taxco is an historic town set in the upland **plateau** south west of Mexico City. Tourism and the sale of silver jewellery are the main sources of income

△ **Figure 4.41** Traffic congestion is a major problem but there are plenty of taxis

Assessment tasks

△ **Figure 4.42** The old village shop, Hilderstone, 100 years ago

△ **Figure 4.43** The building today

Hilderstone is a village in Staffordshire, England (see pages 68–71). Like most villages it began hundreds of years ago. The village then was the centre of life for local people. Look at Figures 4.42 and 4.43. How has village life changed?

Recently, new houses have been built in this village (Figure 4.45). Others could be bought and improved (Figure 4.44). Some people are concerned about whether more new houses should be built. A special Parish Council meeting has been called to discuss the matter.

△ **Figure 4.44** Housing that could be modernised

△ **Figure 4.45** New housing

△ **Figure 4.46** Old people's bungalows

△ **Figure 4.47** The Roebuck pub

There are five main options:

A Build 20 starter homes on the edge of the village for local people on low income.

B Allow old buildings and barns to be converted. They will need to fit in with the style of the village.

C Build 25 old people's bungalows for local people.

D Build 10 executive style homes. They will have double garages and five bedrooms each.

E Do nothing! The village should not expand.

Five members of the public have turned up. They are:

- A young married couple with two toddlers. They are on a low income.
- The owner of the pub 'The Roebuck'.
- A manager of a pottery firm in Stoke-on-Trent who is looking for a large family house and would like to live in this village.
- A farmer who owns land on the edge of the village.
- An unemployed labourer, living in the village.

Activities

1 Draw a copy of this table and fill it in.

Option for village	Good point	Bad point	People at the meeting who will like it
A Starter home			
B Old buildings converted			
C Old people's bungalows			
D Executive homes			
E Do nothing			

2 What do you think should happen, and why?

Assessment tasks

Report writing

This chapter has looked at three places: Hilderstone, Mexico City and Cardiff Bay.

Your task is to report on one of these places for a Year 6 primary school class.

Think about how you are going to present the information. It can be a poster, a leaflet or a booklet.

Plan the questions you are going to report on.
For example:
• Where is it?
• What is it like?

• Who lives here?
• How has the place changed?
• How has this affected people?

You can use the questions as headings for your report.
You should include maps, photographs and text.

Try to use short sentences and make the report attractive. Your information should be clear and well-presented.

Designing a website

ICT activities

1 Use the concept map (Figure 4.48) to plan a website to tell people about a settlement you know well.

▷ **Figure 4.48**
Flow diagram for a village website

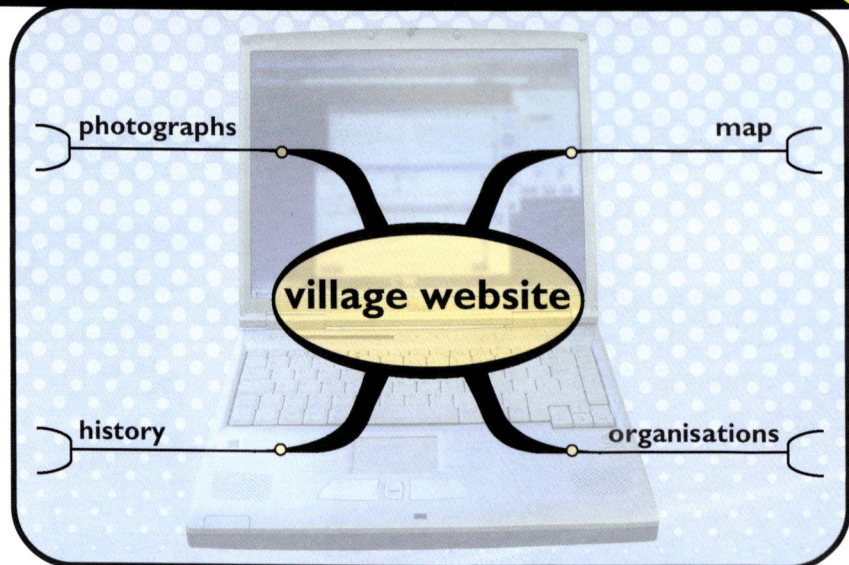

• Is there something to interest children and adults on my website?
• Does my website help people understand what the settlement is like?
• Have I suggested links that people might like to explore after looking at my website?

2 Use the ICT links on page 83 to give you some ideas. There are several sites for villages. Try finding out about a village you know – type its name into a search engine such as **www.google.com**

Using photographs to review your work

One way of finding out whether you have understood this chapter is by choosing two photographs that you *think* are different; for example, one photo could show an urban area, the other, a rural area. Use the compass rose (Figure 4.49) to help you analyse them.

❯ The compass rose

1 Sketch the image in the middle of a sheet of plain paper.
2 Try to describe each photograph under the following headings:

△ **Figure 4.49**

N	**North** stands for Natural. Describe what is natural about the photograph, e.g. vegetation, clouds
S	**South** stands for Social. Describe how people's lives are affected, e.g. where do children play?
E	**East** stands for Economic. Describe how money is made, e.g. what jobs are shown?
W	**West** stands for Who Decides? Describe the decisions made by people that have influenced the photograph, e.g. why a building is here.

Activities

3 What have you found that is similar about the two photos?

Web address
www.hilderstonevil.freeuk.com/
www.old-maps.co.uk/
www.ordnancesurvey.co.uk/getamap
www.local-transport.dtlr.gov.uk/ruralbuses/

What it shows
Shows information about Hilderstone
Finds an old map of your village
Finds an up-to-date map
Contains information on the government plans for buses including rural areas

5 Ecosystems

How are you linked to a blade of grass?

Ecosystems are about survival. Every living thing has a role in an ecosystem. What does each living thing do? Who eats whom? Some links are clear, but other links are less clear.

Most likely to eat or be eaten?

▷ **Figure 5.1** Who eats whom?

What are ecosystems?

Living organisms (**biotic**) and non-living things (**abiotic**) make up an ecosystem. The various organisms depend on each other and the non-living parts to survive. Understanding these links is an important part of studying ecosystems.

Ecosystems are not all the same size. They can cover a large area, like a rainforest, or a small area, like a garden pond.

The links between parts of ecosystems can be very complicated. This makes it difficult to predict how human activity or natural events will affect ecosystems. Change can cause long-lasting effects in the ecosystem, like species extinction.

What has been said about the natural world

'We mention nature and forget ourselves in it: we ourselves are nature'

Friedrich Nietzsche (1844–1900)

'Nature is the symbol of the spirit'

Emerson (1803–1882)

'Extinct is forever'

Friends of Animals

'He who plants a tree, plants hope'

Lucy Larcom (1826–1893) U.S. poet

△ **Figure 5.2**

The need for an organism to survive in an ecosystem means it has to **evolve**. Species change to survive by adapting to take advantage of the environment. Only those organisms that change survive.

Humans have evolved to survive in a variety of ecosystems. We have successfully learnt how to use the resources in different ecosystems.

Different species can successfully use the same ecosystem (see Figures 5.3 and 5.4). The leaf cutter ants and humans are both doing the same thing: using the tropical rainforest's natural resources. However, each affects the ecosystem they are using differently.

This chapter will look at different ecosystems. It will also look at how humans are using resources and threatening these natural environments. It will be important to think about the balance between using ecosystems and not destroying them by overuse.

△ **Figure 5.3** Leaf cutter ants

△ **Figure 5.4** Humans taking resources from the rainforest, Brazil

Activities

1 Copy Figure 5.1. Draw lines to link who eats whom.

2 Suggest what would happen if the one being eaten were removed.

3 Which is the most important organism? Suggest a reason why.

ICT links

There are many great resources on the Internet. Try to do your own research using the following sites as a starting point:

www.nationalgeographic.com/earthpulse

http://mbgnet.mobot.org/index.htm

What do you already know about ecosystems?

Figure 5.5 shows the main parts or **components** of a large-scale ecosystem. It is easy to show the links between parts. However, we must be able to explain these links.

An ecosystem can be broken down into key parts. This is called system thinking. It helps to explain what is happening in an ecosystem.

Inputs	what goes in
Flows	where energy goes to
Processes	what is happening
Stores	where energy is kept
Outputs	the changes
Feedback	how energy returns to the ecosystem

▽ **Figure 5.5** Concept map

Activities

1 Copy Figure 5.5 on a large sheet of paper. This is a concept map.
a Add these features in an appropriate place on your copy: weather, insects, fungus, trees, humans, fox, rabbits, air, birds, cattle and leaves.
b Now draw links between these features.
c Can you add three other features and links?
d Suggest a reason for each link you draw.

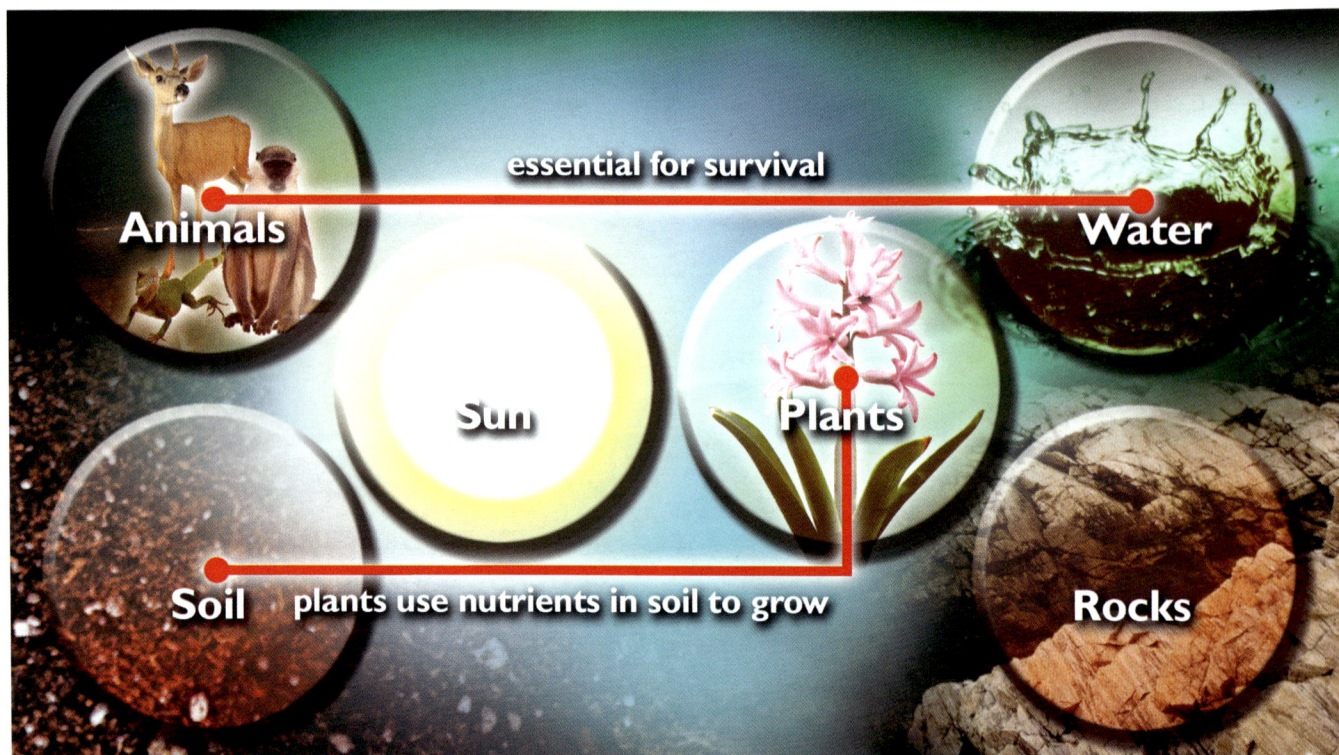

Animals — essential for survival — Water

Sun   Plants

Soil   plants use nutrients in soil to grow   Rocks

What is the point of studying ecosystems?

An important question is how do humans fit into ecosystems. Some believe that we can use the natural world as we see fit. Others believe that humans should take care of our planet for future generations.

Sustainable development is a way to protect the resources of our planet for the future. It supports development that does not destroy the environment forever. When you fell a tree, you would plant one to replace it. However, this can be done for some resources but not for others.

We use oil, gas, coal and nuclear power to produce electricity for our hi-tech society. Once used, these resources cannot be replaced. Burning these fuels causes pollution and damages the environment. Alternative energy sources like solar power, wind and hydroelectric generators are a sustainable resource. These are limitless and clean resources.

▷ **Figure 5.6** Examples of sustainable use of resources: (i) wind farm, (ii) replanting forests, (iii) re-seeding a flower meadow, (iv) solar panels on a house

Activities

2 Write the words below on separate pieces of paper. Use a dictionary to find their meanings.

Exploitation	Industry
Energy	Earning a living
Politics	Sustainable
Destruction	Extinction
Research	Replacing

Sort the words into two piles.

1 Words that can be linked to looking after ecosystems.
2 Words that can be linked to destroying ecosystems.

Explain how you decided on your two piles.

ICT links

To find out more about alternative energy supplies and sustainable development, follow these links:

The Centre for Sustainable Energy
www.cse.org.uk

The World Wildlife Fund, Sustainable Development site
www.wwflearning.co.uk/ourworld

What is the global distribution of ecosystems like?

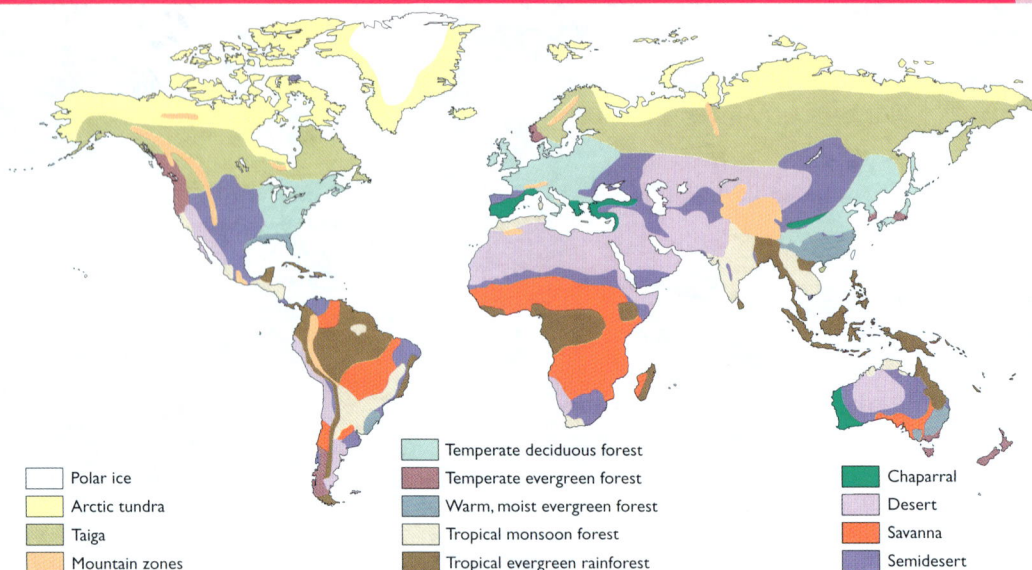

Key:
- Polar ice
- Arctic tundra
- Taiga
- Mountain zones
- Temperate deciduous forest
- Temperate evergreen forest
- Warm, moist evergreen forest
- Tropical monsoon forest
- Tropical evergreen rainforest
- Chaparral
- Desert
- Savanna
- Semidesert

△ **Figure 5.7** Distribution map

Light, heat and moisture create the conditions for all life on Earth. They determine the location of different ecosystems. Figure 5.7 shows the distribution of different ecosystems. These large-scale ecosystems are called **biomes**.

Biomes have developed in areas with similar conditions for life. There is a strong link between climate and distribution of the world's biomes.

Figure 5.8 shows how the Sun's energy provides the Earth with light and heat. The heat evaporates moisture from the Earth's surface. This creates an atmosphere. The Sun's energy unevenly heats the Earth's surface. The Earth gets cooler away from the equator. The Earth's curve means that the heat of the Sun is less concentrated at the poles than at the equator.

As the Earth orbits the Sun, the poles tilt towards and away from the Sun. This causes seasons away from the equator. This means the length of daylight and heat will vary, and with it the growing season for all plants.

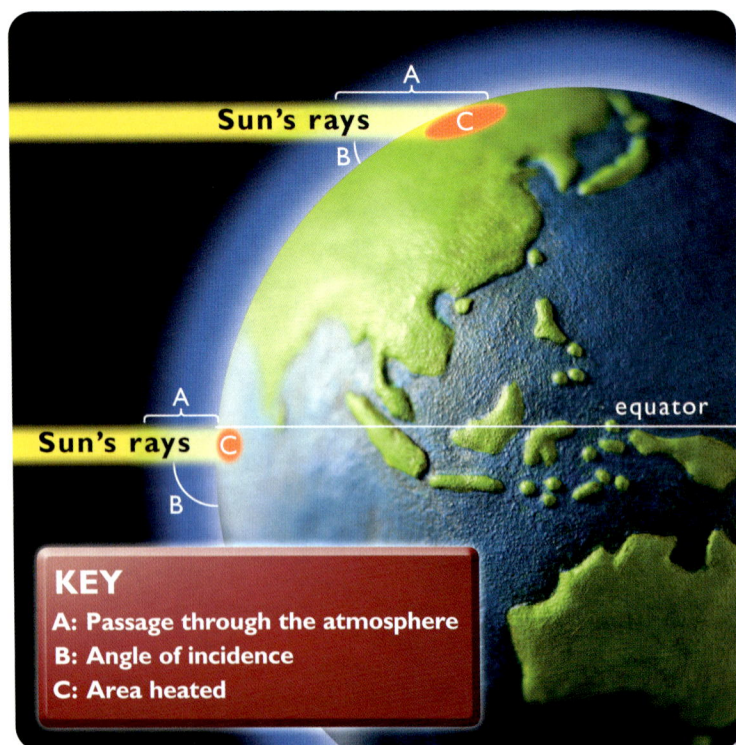

KEY
A: Passage through the atmosphere
B: Angle of incidence
C: Area heated

◁ **Figure 5.8** How the Sun's rays affect the Earth

Why do ecosystems change?

Plants and animals can adapt to their environment. This has enabled a wide range of life to evolve. Each ecosystem has developed a unique range of living things.

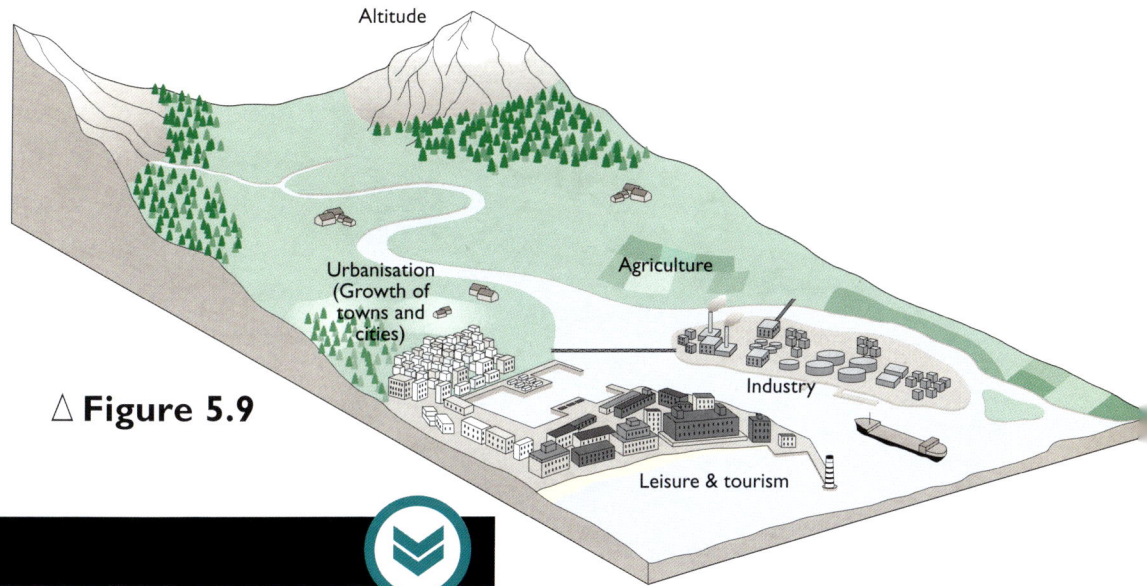

△ **Figure 5.9**

Activities

As living things evolve, they adapt to live in different conditions. It is possible to have closely related species living in very different ecosystems. But why should living things adapt?

This activity will help you to think about evolution in a slightly different way! You are going to draw a new animal.

1 Choose one of these animals:
• A **herbivore** (plant eater) living in a hot climate
or
• A **carnivore** (meat eater) living in a very cold climate

In the box below, some useful ideas are listed. You can add others.

Draw what you think this animal would look like. Label the adaptations it would need to survive in its ecosystem.

Surviving the climate	Finding food	Avoiding trouble
Thick fur coat	Fast runner	Camouflage
Active at night	Sharp claws and teeth	Smells bad
Doesn't need much water	Good sense of smell	Lives in large groups

2 Display your ideas set against a drawing of the landscape of a hot or cold ecosystem.

However, ecosystems do change. If this change is rapid, many living things find it difficult to adapt to the new conditions. When this happens they can become extinct.

Human activity has affected most ecosystems in some way. People do not always get the balance right! The actions of humans have caused many species of plants and animals to become extinct.

ICT Activity

Using a search engine like:
www.google.com
www.lycos.com
www.yahoo.com
find out more about these extinct species:
• Mammoth
• Dodo
• Giant Sloth

How are ecosystems organised?

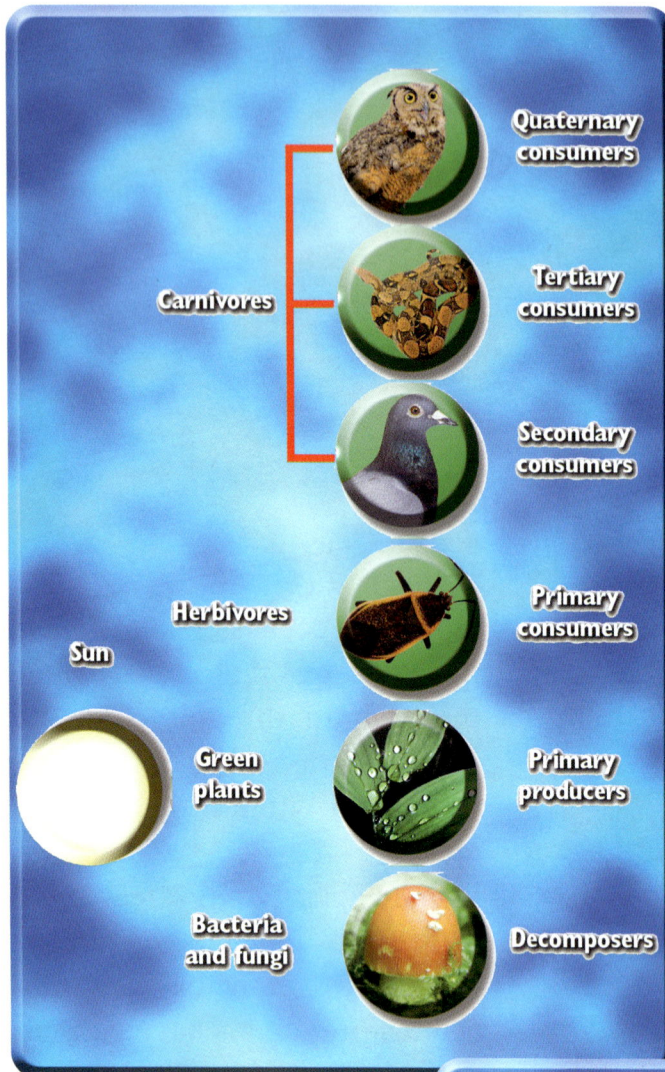

△ **Figure 5.10** A food chain

The parts of an ecosystem seem simply linked. However, when we look at the whole ecosystem, these links can be complex. By looking at how energy (food) moves through an ecosystem, it is possible to understand these links. By putting the different parts into categories it is possible to see how the ecosystem works.

As energy passes up the **food chain**, fewer organisms use it (Figure 5.10). At the top of the food chain, the organisms are more complex **consumers**.

A **food web** shows the links between different organisms in a food chain (see Figure 5.11). There are often many links between different organisms.

The food web shows how ecosystems need all their parts to survive. If change damages the food web, the ecosystem could be at risk of collapsing.

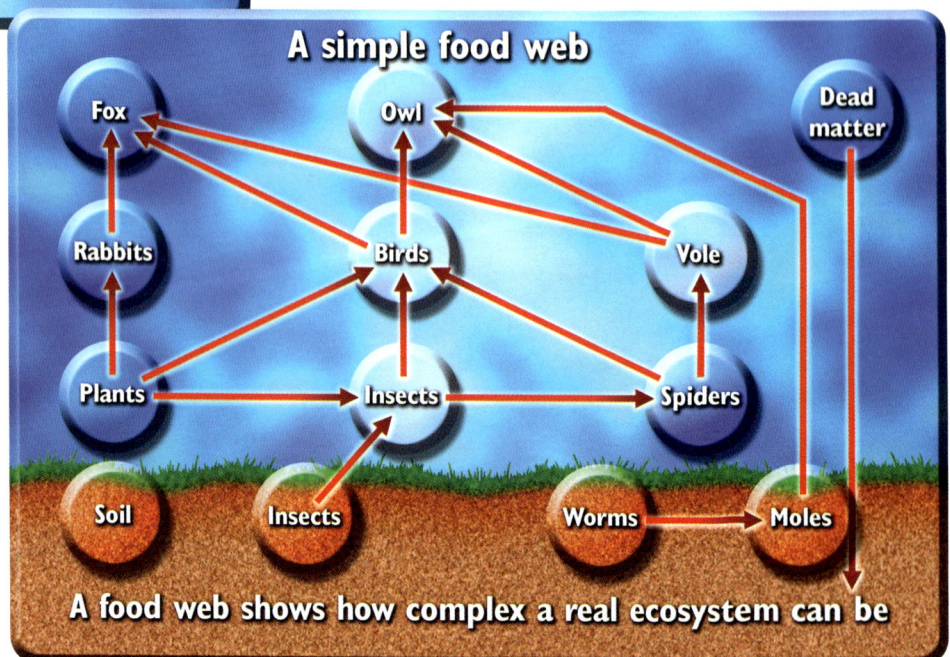

▷ **Figure 5.11** A simple food web

Cyanide Spill Floods into Danube

adapted from article by Nick Thorpe in Budapest, Monday 14 February, 2000 for the *Guardian*

Yesterday a 40 km spill of toxic cyanide washed down the River Tisza, northern Yugoslavia. A tide of dead fish followed the flow of poison downstream into the River Danube. The spill came from the Baia Mare gold mine in Romania. A dam overflowed and cyanide flowed into the streams.

Biologists fear that the River Tisza may never recover. Others think it will take 12 years to recover from the poison.

Fishermen set nets to catch the dead fish. They collected 300 tonnes from the Tisza.

The Mayor of Senta, Yugoslavia, estimated that 80% of the River Tisza's fish had died.

If new fish were put into the river, they would die because there is no food. The poison has destroyed the food chain.

Officials have demanded compensation from the Australian mine owners. The company has played down the disaster. They admit there has been a significant overflow from the dam, but say this is not an ecological catastrophe.

△ **Figure 5.12**

Activities

1. Use Figure 5.11 as a guide to draw a food web for a river ecosystem.

2. Could any part of your food web survive without the producers?

3. Read Figure 5.12 with your teacher. Make two lists from the news report.
   - Three facts about the incident.
   - Two different opinions of people involved.

4. Suggest how the cyanide poison has damaged the river ecosystem.

5. Who is responsible for this incident? Suggest what they should do to save the river ecosystem.

Tropical rainforests: A biome under threat

A warm, moist wind blew from the south down a valley lost in green folds of undulating rainforest. Carried on the wind were thick, earthy smells of sweet-scented plants and decomposing leaf litter that carpeted the jungle floor. This enticing fragrance conveyed the first hints of what life would be like within the 120 000 square miles of rainforest that separated me from the coast by more than 400 miles.

Rainforest biomes cover large areas of the world. These areas occur between latitudes 10° N and 10° S of the equator.

◁ **Figure 5.13** From *Stranger In The Forest* by Eric Hansen

▷ **Figure 5.14** World map of forests

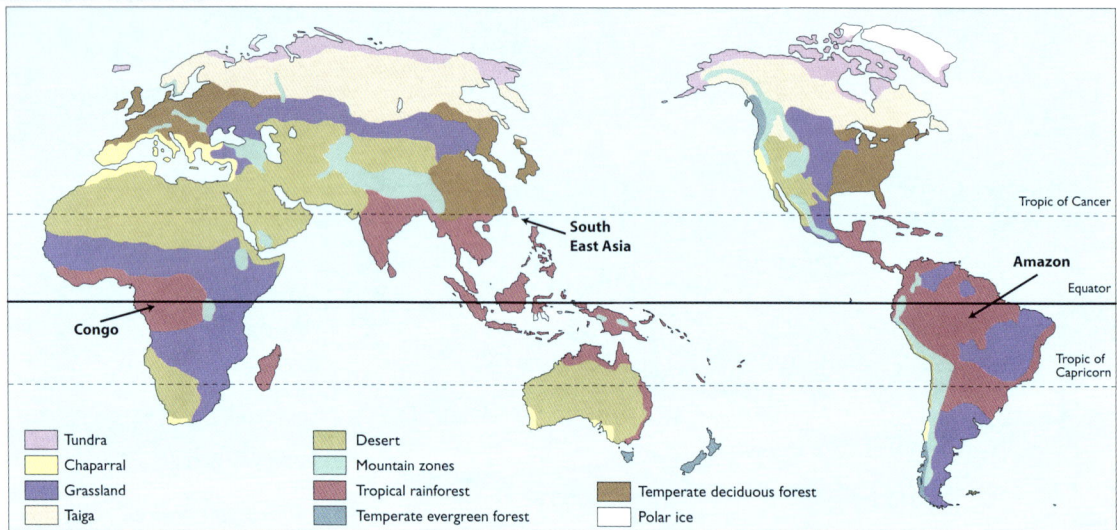

Legend:
- Tundra
- Chaparral
- Grassland
- Taiga
- Desert
- Mountain zones
- Tropical rainforest
- Temperate evergreen forest
- Temperate deciduous forest
- Polar ice

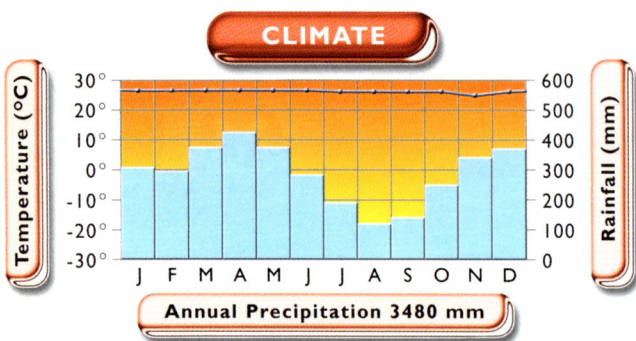

CLIMATE

Annual Precipitation 3480 mm

△ **Figure 5.15** Climate graph for the rainforest region

The tropical rainforest climate is the same throughout the year. It is always hot! It rains most days with heavy downpours of rain. The hot, moist conditions create clouds during the day that burst by the late afternoon.

This hot, wet climate produces **vegetation** that grows quickly. It also grows big! Imagine trees 10 times the height of your house, and leaves large enough to cover your bed.

There are no seasons so the plants grow all year. The lush vegetation supports a huge and unique variety of animals and insects.

The tropical rainforests of the world are a very important part of the global ecosystem.

◁ **Figure 5.16** In a tropical rainforest

△ Figure 5.17

EMERGENT LAYER
- The tallest trees called Emergents.
- The most sunlight.
- Eagles, monkeys and bats live here.

CANOPY LAYER
- Trees about 30–40 metres high.
- Average sunlight.
- Most wildlife: birds, monkeys, tree frogs and snakes.

UNDERSTOREY LAYER
- Plants have large leaves to catch what sunlight they can.
- Average sunlight.
- Animals include big cats like jaguar and leopards. Also many species of insect.

FOREST FLOOR
- Very little grows on the forest floor as it is so dark.
- Least sunlight.
- Ground-living animals like giant anteaters, Capybara and wild boars live here. Lots of insects and fungus.

Activities

1. Close your eyes. Have a partner read Figure 5.13 aloud to you. Imagine the sounds and sights of a rainforest. When they finish reading, keep your eyes shut for another minute. Keep a mental picture of the rainforest in your mind.

2. With your partner, draw what you think a rainforest might be like.

3. Use the ideas from pages 92–3 to add another 10 ideas about rainforests.

Structure of the ecosystem

The forest has a chaotic, unorganised appearance. However, there is a well-defined structure (Figure 5.17). There are a series of layers at different heights above the forest floor. This structure depends on how the plants compete to gain the most sunlight.

The top canopy layer has the most life. Here the dense mass of branches and leaves get most sunlight. The fruit of the trees is abundant and attracts animal and bird life. The amount of sunlight gets less the closer you are to the forest floor. The forest floor is very dark. The floor is where the dead and decaying material ends up. Organisms like fungi and termites break this material down and help to recycle these valuable nutrients.

Nutrient cycling

The warm, moist climate conditions mean the rate of plant growth and decay is faster than in other biomes. Nutrients produced by the decomposing organisms provide food for plants to grow. Plants get most of their nutrients from this source. The rainforest has poor soil that has few nutrients.

How are the rainforests under threat?

Reasons for exploitation

The growing conditions of the tropical rainforest are ideal for vegetation. These areas contain large amounts of resources of timber, metal ores and other rainforest products. There is a growing worldwide demand for these products. Rainforest countries, like Brazil, are under pressure to exploit these resources. The dots in Figure 5.18 show other rainforest areas.

Nobody knows whether the demand for rainforest products will slow or stop. However, we know that the area of rainforest not affected by human exploitation is rapidly shrinking.

This is a global problem (see Figure 5.18).

▽ **Figure 5.18a** Timber extraction for hardwood furniture in Balikpapar, Indonesia

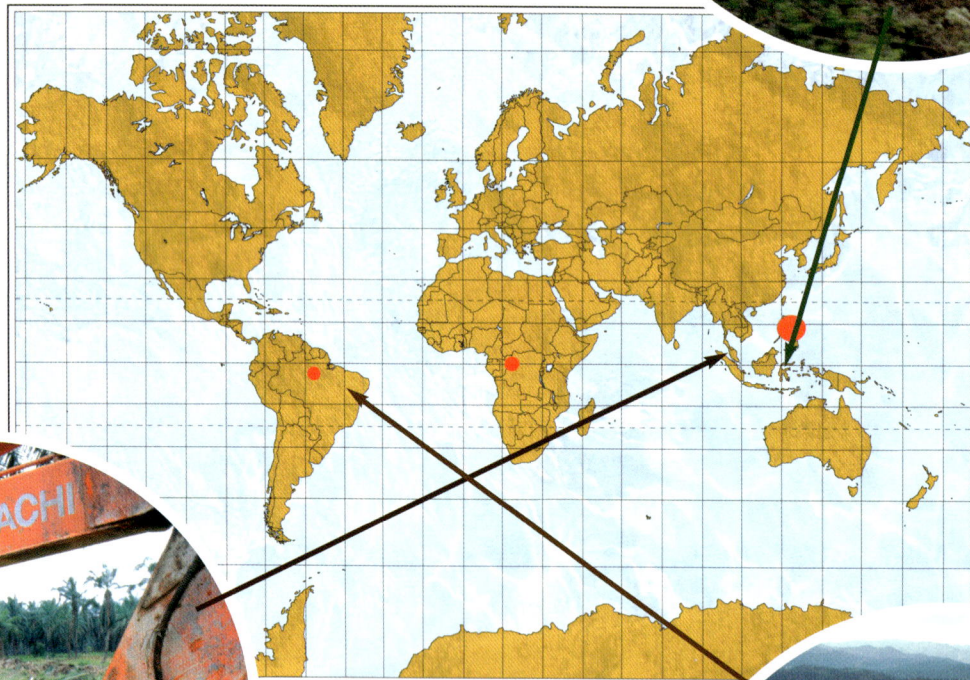

△ **Figure 5.18**

◁ **Figure 5.18b** Extensive oil palm plantation in Malaysia

▷ **Figure 5.18c** Open-cast mining

What?	Why?	Who benefits?	Effect on the forest
Medical research	Many rainforest plants provide chemicals to make useful medicines	MEDC scientists	Need to conserve these known and yet-to-be-discovered plants
Timber	For exotic hardwood timber, plywood and paper making	Exports to MEDCs	Global deforestation. 50% of timber and 75% of paper production is used in the MEDCs
Oil palm plantations	Production of vegetable oil for export	LEDCs earn money. Malaysia is the world's leading producer and exports 85% of its production	Forests replaced by plantations
Cattle ranching	Keeping beef cattle for meat exports	MEDCs import cheap meat	Ranchers slash and burn large areas. One square km of forest produces 181 burgers. The land soon becomes useless
Mining and quarrying	Extracting valuable ores like gold and tin	MEDCs own the companies	Large open-cast mines. Chemicals used to separate the ore from the rock. Forests are destroyed

Activities

Study pages 94–5.

1 What is the overall effect of humans exploiting the rainforest?

2 Would you buy a gold bracelet if you knew it came from an area destroyed by mining? Suggest a reason for your answer.

3 Who seems to benefit most from exploiting the rainforest? Is this fair?

4 Should there be a ban on people exploiting the rainforest? Suggest a reason for your answer.

How can an ecosystem be managed?

There are very few ecosystems that people have not affected. When people first see an area, their first instinct is to find out how they can benefit from its resources. These may be obvious, like timber from rainforest trees, or less so, like an underground oil reserve. However, we cannot actually touch some resources. A spectacular view from the top of a hill, or the experience of walking on a beach, are just as valuable.

In the last few years, people have begun to think about the damage being done to ecosystems. People who wish to use an ecosystem's resources are more aware of the damage. They will limit the damage done by some form of management.

Aims of management

Managing ecosystems is all about balance. The balance is between how to use the resources and how to make sure that any damage done is minimal. Management aims to protect and conserve the ecosystem.

Ecosystems provide different resources for different people. Sometimes there may be **conflict** in using the resources. Successful management must plan how different groups can use the ecosystem resources. If conflict occurs, it should be resolved through management.

Conflict resolution

Conflict arises when two or more groups want the same resource at the same time. It also occurs if one group activity affects another group. Think about your family. How many times have they clashed over using the TV, telephone or bathroom?

One way to resolve conflict is to allow different groups access to the same resource at different times. Zoning helps by restricting the use of a resource to a specific area and group. There are other ways to manage conflict: can you think of any?

What is management?

We use the word management a lot in geography, often in connection with many different topics, but what does it actually mean?

In most situations it is used to describe **the control of a variety of factors to achieve a specific purpose.**

What is sustainable management?

To use the resources of an ecosystem so that it is maintained.

▽ **Figure 5.19**
Conservation area

Buffer zone management

This is a way of controlling access to sensitive areas. Buffer zones are widely used around the world and for different situations. A famous example is the rainforest of the Korup National Park, Cameroon, Africa. This management effectively stops ecological damage.

The Core Area
No access to the core is permitted. There is as little human contact with the core as possible. Some scientific monitoring may be allowed.

The Transition Zone
While there are no restrictions on the number or type of activity that happen here, all potential threats are monitored closely.

Transition Zone
Buffer zone
Core Area

△ **Figure 5.20a**

The Buffer Zone
Limited access is allowed for certain activities; for example, agriculture, recreation, education and scientific research. People's movements are tightly controlled with penalties for those who ignore the rules.

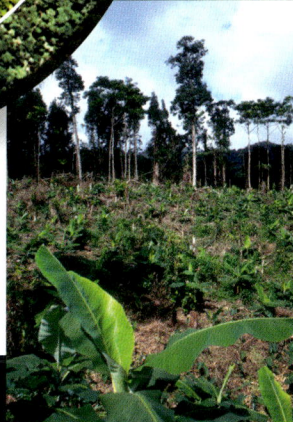

△ **Figure 5.20b**

◁ **Figure 5.20c**

Activities

1. Draw a 1 cm border around a piece of plain A4 paper.

2. The rectangle you have drawn represents an area of rainforest. The scale is 4 cm = 1 km.

3. Add a road that goes from top left to bottom right across the map.

4. Add a river that crosses from bottom left to top right across the map.

5. Before planning the use of this area, you must identify a core area of at least 25% of the area. This must remain untouched and conserved as primary forest.

6. On your map, decide where to locate:
   • two hotels with road access
   • an area for visitors to trek in the forest
   • three villages with farmland to supply food to the hotels (these need a 2 cm circle)
   • new roads or tracks for access.

7. Write a brief report on your map. Are there any conflicts? What was most difficult to fit in? How well did you manage the use of resources? Compare your plan with others in your class.

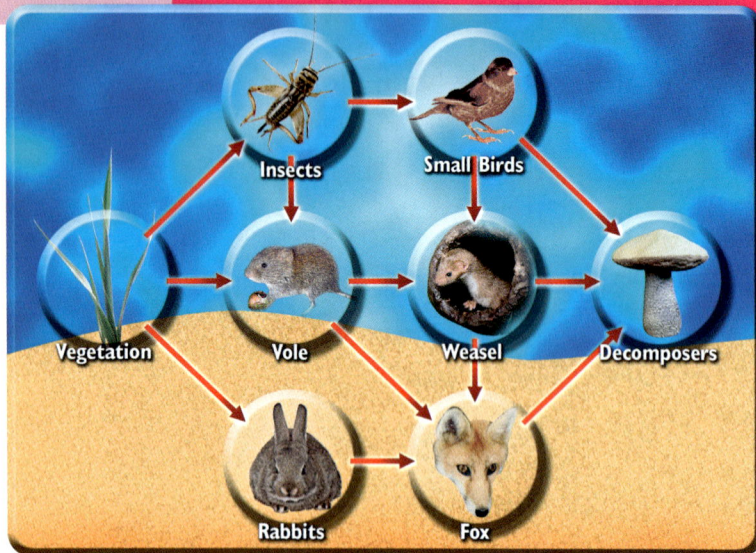

△ **Figure 5.21** The sand dune food web

Coastal environments feature a variety of ecosystems. One of the most interesting is the sand dune ecosystem.

Sand dunes occur along the north-east coast of England. Druridge Bay is a part of this impressive coastline. It stretches from Hauxley in the north to Creswell in the south. UK, European and international laws protect the area and the sand dune system (Figure 5.22).

The Country Park, at 2600, is popular throughout the year with visitors. The nature reserves, 2802 and 2796, attract people wishing to experience the sand dunes.

Druridge Bay is an example of good ecosystem management. However, what do they do?

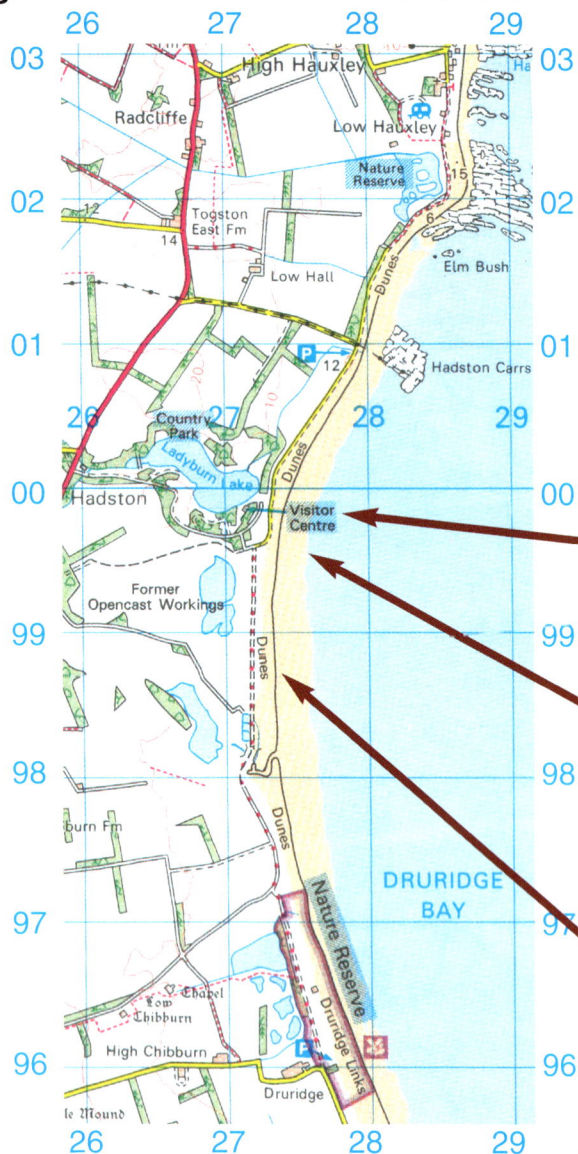

The visitor centre has a large car and coach park. It has toilets and a small café. The centre is an information point with public display boards and information leaflets about the sand dune ecology and management activities.

In front of the visitor centre, the dunes have a high level of management. Paths direct people to the sea and fences prevent access to sensitive areas.

Further south, the dunes have less management. Fewer people use these dunes. There are a number of informal paths. There are no access restrictions on people.

© Crown copyright 2002

◁ **Figure 5.22** Map of Druridge Bay

△ **Figure 5.23**

People visit the sand dunes for different reasons (see Figure 5.23). However, visitors can cause problems to the ecosystem.

One way to prevent problems is to control people's access. Fences, steps and boardwalks direct visitors along certain paths. Advice guides people about what they can do.

Management to protect the dunes

△ **Figure 5.24** Fences and steps

△ **Figure 5.25** Board walks and paths

△ **Figure 5.26** Visitor centre and car park

Activities

1 Study Figure 5.21. How could visitors to the dunes affect this food web?

2 Make a copy of Figure 5.23.
 a Add three more threats.
b Around the outside add five ideas to prevent these threats.

3 Study Figures 5.24 and 5.25.
 a Draw each picture.

b Add at least two labels to describe how the dunes are being protected from visitors.

4 Study Figure 5.22.
 a Give the grid reference for the Former Opencast Workings.
b How would the reopening of this coal mine threaten the dunes?
c Would you let this happen? Suggest a reason for your answer.

Coral reefs are described as 'Essential Life Support Systems' (World Conservation Strategy 1980). They provide food, health and other aspects of human survival and sustainable development.

ICT links

www.fisheyeview.com

www.cyberlearn.com

△ **Figure 5.27** World distribution of coral reefs

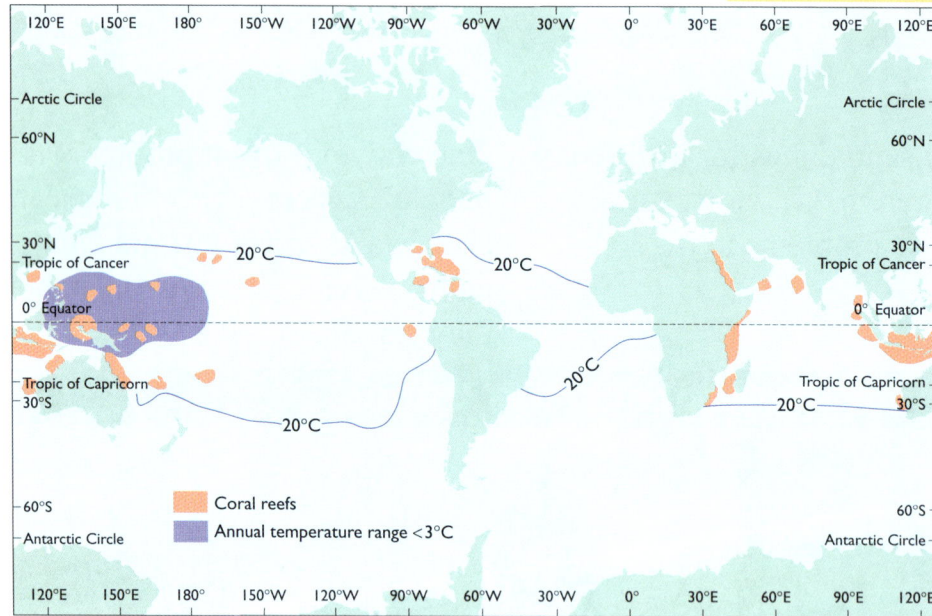

△ **Figure 5.28**
A coral reef

What are reefs like?

Coral reefs grow in tropical seas. Coral reefs grow near 100 countries. These countries lie between the Tropics of Cancer and Capricorn. The total global area of live coral is 600 000 km².

Coral is made from the skeletons of tiny organisms called polyps. Coral grows in warm, clear, shallow seawater. Some coral can grow quickly, but most grows very slowly to form ancient reefs. Reefs are home to thousands of species of marine life.

Coral reefs are very old, rare and sensitive ecosystems. They cannot survive great changes in their environmental conditions.

Pressure on reefs

As coral reefs grow in tropical seas, they are often near popular tourist locations. They are a main attraction in these areas. They attract people who want to scuba dive and snorkel. Local people use the reefs as fishing grounds and as a source of ornaments for tourists.

The coral reefs of Jamaica

Jamaica is a Caribbean island. It attracts a large number of tourists who wish to explore its coral reefs. The demand for more tourist hotels threatens over-use of the coral ecosystems. This could lead to permanent damage of these unique ecosystems.

Assessment tasks

Sunshine Beach Project, Montego Bay, Jamaica, W.I.

Dear Sir/Madam

I am pleased to tell you that the 'Sunshine Beach Hotel' development near Montego Bay, Jamaica, is nearly ready to begin.

The Environmental Risk Assessment will tell us of any likely problems the hotel may cause to the local ecosystem. This development will begin once the project has been assessed.

This hotel will:

- have 500 bed spaces and will be full for nearly all the year

- have a new marina giving access to the hotel. This will attract more visitors and will moor 100 boats. However, a new channel must be cut through the reef

- offer a wide range of water sports like jet skiing and scuba diving

- benefit local people from the jobs in the new hotel and from the tourist trade.

Yours faithfully,

Sarah King

Sarah King
Manager – The Sunshine Beach Project

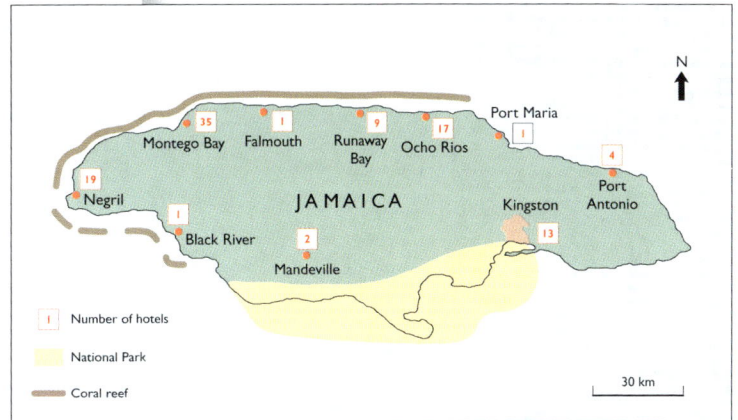

△ **Figure 5.30** Location of main tourist areas

▷ **Figure 5.31**

▽ **Figure 5.32**

Impacts on:	Positive	Negative
People	1. Local people will get jobs building the hotel and marina. 2. Local people will get jobs working in the hotel. 3. Local people will find jobs in the tourist industry.	1. The hotel profits will go back to the MEDC country that owns the hotel. 2. The local jobs are low paid and seasonal.
The ecosystem	**?**	1. More tourists to the reef will damage the coral. 2. The new channel will damage the reef.

Assessment tasks

You are a local newspaper reporter. You are finding out about how tourism is affecting Jamaica's coral reefs. You have collected the following pieces of information. You are aware of the threats to the coral reefs. You need to read each person's statement carefully to make sense of what you have found out.

Frederick has high hopes for his illegal business operations.

Frederick remembers his grandfather's stories of fishing the coral reefs for food. There is high unemployment in the area.

Silt and mud blocks sunlight from the polyps, causing them to die.

Sergeant McDougal has just locked up two fishermen who used dynamite on the coral reef west of an island near Negril. This has been illegal for the past three years.

James and Tina Douglas book a holiday on the Caribbean island of Jamaica. They want to spend their time relaxing in the sun.

Too many boats at the reef today meant the water was cloudy with sediment. Ron's boat had to return early as divers could not see the coral.

Local people take pieces of coral to sell or make into jewellery. This is against the law, but difficult to prevent.

Ron Johnson has just bought three new boats for his 'Coral Reef Boat Tours' Business.

Coral reefs are very sensitive to changes in light, temperature, water clarity, salinity and oxygen.

Montego Bay power station will have to be upgraded to meet the growing tourist demand. Even more warm water will flow from the cooling system onto the reef.

Local hotels have just increased their order for John's lettuces. He will have to contact his fertiliser supplier again.

John's bank agrees a loan for him to buy land by the river for a business extension.

The planned sewer system may smell. This will cause complaints from the residents. Arthur, the site manager, wants a longer pipe to take the waste further out to the sea by the reef.

Task

1 Make a large copy of Figure 5.32 on page 101. Look at the evidence. Decide which box each piece of evidence should go in. Write this on your copy.

2 Do you think that the hotel is a good or bad idea for this area? Suggest a reason for your answer.

3 Design a newspaper page. Write a 50-word article for your newspaper explaining why the new hotel is a good or bad idea. You could add some pictures to support your ideas.

◁ **Figure 5.33**

Review

This chapter has looked at ecosystems and how they are affected by people.

This review page is to help check on your understanding of this topic. Start with the first column and read the statement in the bottom box. If you agree with it, move on to the box above. To show your understanding at each level, write an 'I can …' sentence by changing the statements from a question.

If you want to move up from one box to another but are unsure, ask your teacher for help.

› Issues to investigate further

• The Eden Project
• Overfishing in the North Sea
• **Eutrophication** – Wetlands' water quality and agriculture
• The damage done by holiday developments in less economically developed countries (LEDCs).

Can you explain how change to one part of an ecosystem can affect and be traced through the whole system?	Can you explain the location of more than one global ecosystem?	Can you explain the idea of energy flows within food chains and how damaging the food chain can affect different organisms?	Can you explain the effects of exploitation on the ecosystems?	Can you give examples of where sustainable management has helped conserve the ecosystem's resources yet allows people to use them?
Can you name and describe how biotic and abiotic parts link in an ecosystem?	Can you explain what biomes are and how their location is related to the amount of energy they get from the Sun?	Can you explain how organisms form complex food webs within ecosystems?	Can you explain why people exploit the natural resources of a named ecosystem?	Can you explain what sustainable management is and why it is important where humans threaten ecosystems?
Can you describe an ecosystem as its inputs, flows, stores and outputs?	Can you describe how temperature and precipitation affect the growth of plants and animals in ecosystems?	Can you give an example of a food chain for an ecosystem?	Can you describe what exploitation is, giving an example of an ecosystem where this has happened?	Can you describe what ecosystem management is, giving examples of where it has been used?

6 The Environment

Is this a tale of two cities?

Figures 6.1 and 6.2 are of Longton. It is the same view but 100 years apart. The church tower is in both photographs. However, how has the view changed in the last 100 years?

▽ **Figure 6.1** Longton 1900

▽ **Figure 6.2** Longton today

Activities

⌄

1 Study Figures 6.1 and 6.2.
a Suggest at least three changes to the view in 1900 and today.
b How many differences did your class find?

2 Imagine that you were back in 1900 Longton.
a What would Longton be like? What smells and noises would there be?
b What would the houses look like? Draw a house from that time.
c What jobs would people have?
d What would surprise you the most about 1900 Longton?

3 Study Figure 6.1.
a What was the environment of Longton like in 1900?
b How might this 1900 environment affect the health of people in Longton?

4 Study Figure 6.3.
a What were the three main causes of death in 1900?
b Which diseases are less common today?

5 Study Figure 6.4. People in 1900 had a shorter life expectancy. People live longer today.
a What shape does the 1900 **population pyramid** have?
b Which population pyramid has more older people?

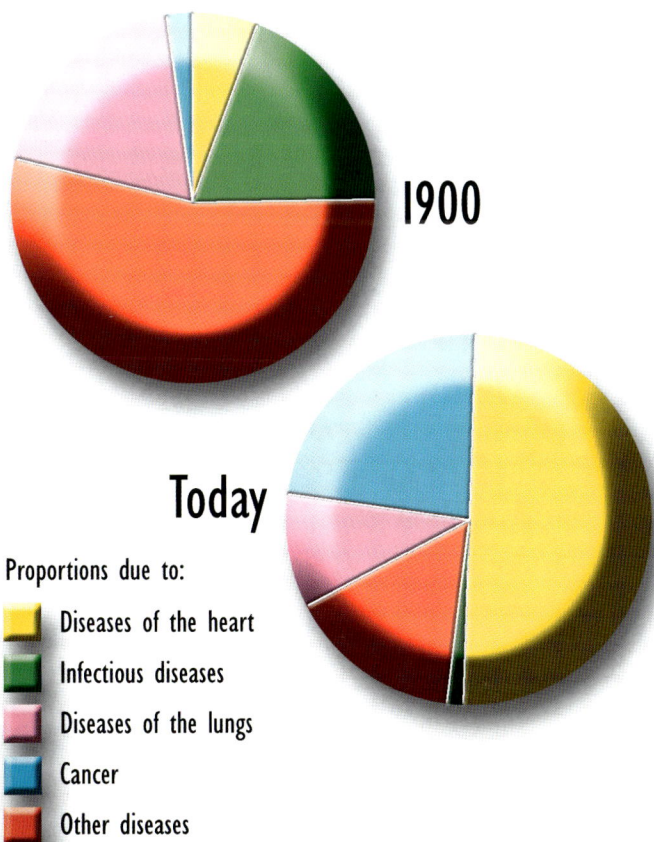

Proportions due to:

- Diseases of the heart
- Infectious diseases
- Diseases of the lungs
- Cancer
- Other diseases

△ **Figure 6.3** Causes of death in Longton, 1900 and today

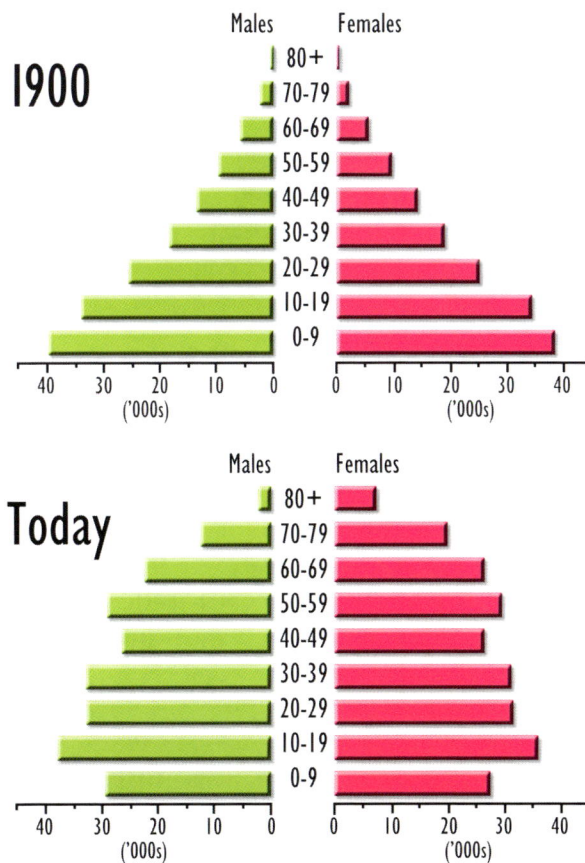

△ **Figure 6.4** Population pyramids for North Staffordshire, the area where Longton is situated, 1900 and today

What is the state of the planet?

△ **Figure 6.5** Earth from space

The **environment** means our surroundings. This includes the air, land and water on our planet. Many people are concerned about the global environment. How we use our resources affects our planet and causes air and water pollution.

Activities

1. Copy Figure 6.6.

2. Label one side rich countries and the other side poor countries.

3. Is this a fair share of resources?

△ **Figure 6.6**

Using up resources causes **pollution** and climate change. Plant and animal species can become **extinct**. We are all to blame for the state of our planet.

On a world scale

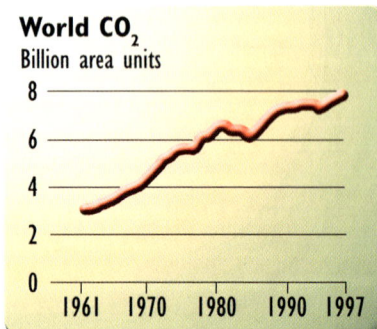

△ **Figure 6.7**

Carbon dioxide is a gas produced when we use **fossil fuels**. Carbon dioxide (CO_2) is a greenhouse gas. This gas traps heat and warms up the Earth. This is **global warming**. Ice caps melt and sea levels rise.

Forests absorb carbon dioxide. To cope with the increase of this gas, the planet needs more trees.

Communities

Rich countries use more resources and produce more waste. One third of the food waste of the USA could feed 26 million people for a year.

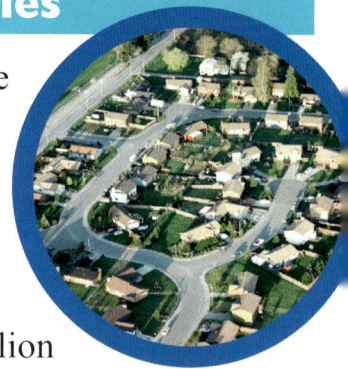

Commerce and industry

Many products have a limited use. The USA throws away 15 million working personal computers a year.

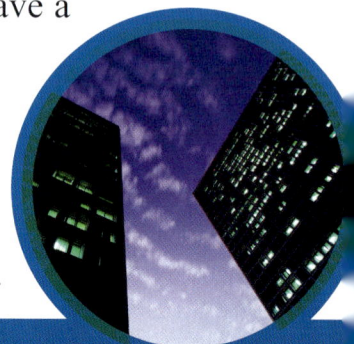

Consumption

We all need to consume to live. Using local resources we would save the energy used in transporting goods over long distances. If Europe produced the 800 000 tonnes of imported wheat used for European bread instead of importing it, there would be a big saving in costs and pollution.

Production

Making products uses up the planet's **raw materials** or **energy**. **Globalisation** encourages a worldwide trade in goods. **Recycling** waste can reduce the pollution. By recycling plastic waste, Britain could reduce carbon dioxide emission by 3 million tonnes a year.

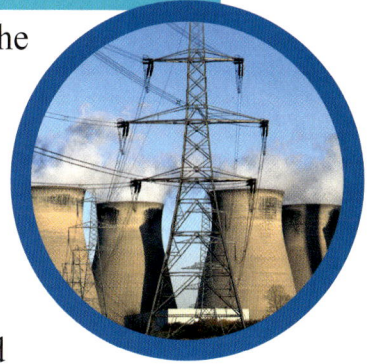

Waste

Each week Britain produces enough rubbish to fill Wembley Stadium. Rich countries produce the most waste. The USA produces 1 410 kg per person each year.

Natural resources

Some resources are **non-renewable**, such as fossil fuels. **Renewable resources**, such as timber, are in decline. Commercial fishing does not let fish stocks recover. Fish stocks are decreasing.

Activities

4 What activities produce carbon dioxide?

5 How will global warming change the lives of people living near the coast?

6 Copy and complete this table.

Cause of waste	Problem	A solution
Communities		
Commerce/Industry		
Consumption		
Waste		
Production		
Natural resources		

7 Draw a poster to show how people can reduce waste.

Is water quality in danger?

The **environment** means our surroundings. This includes the air, land and water on our planet. Many people are concerned about the global environment. How we use our resources affects our planet and causes air and water pollution.

Using up resources causes **pollution** and climate change. Plant and animal species can become

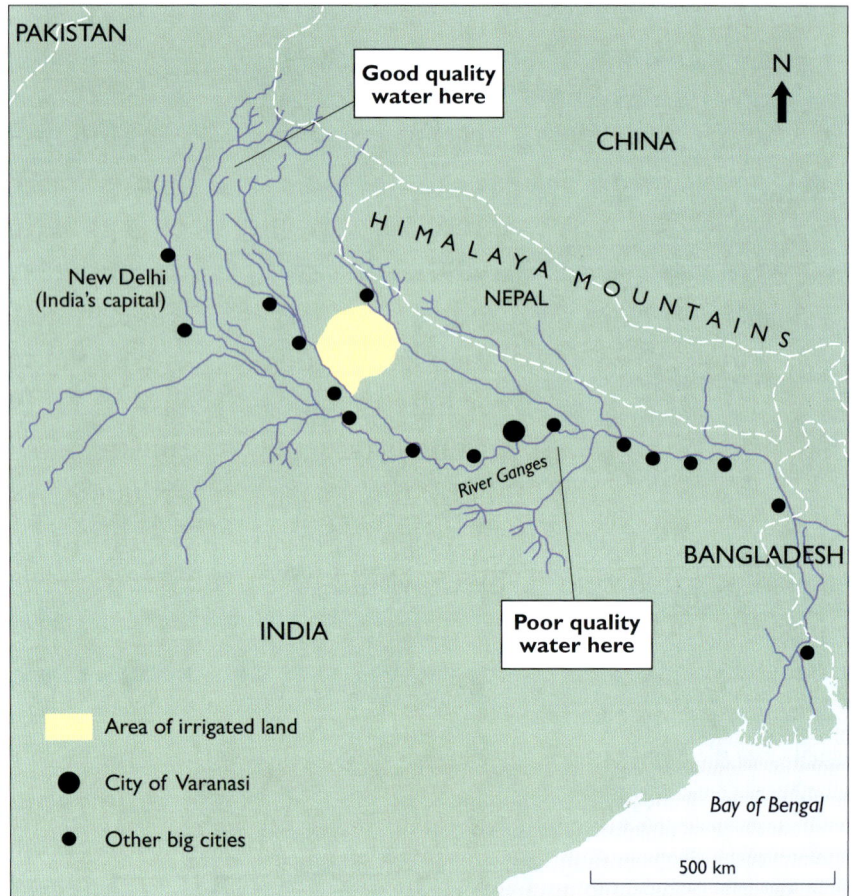

△ **Figure 6.8** Map of the Ganges

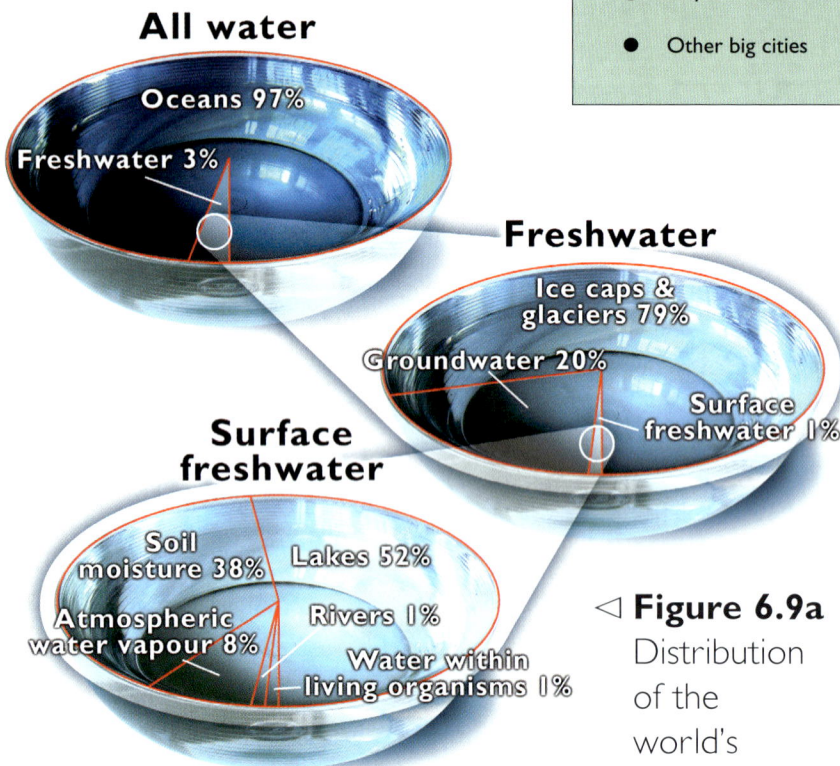

All water

Oceans 97%
Freshwater 3%

Freshwater

Ice caps & glaciers 79%
Groundwater 20%
Surface freshwater 1%

Surface freshwater

Soil moisture 38%
Lakes 52%
Atmospheric water vapour 8%
Rivers 1%
Water within living organisms 1%

◁ **Figure 6.9a** Distribution of the world's water

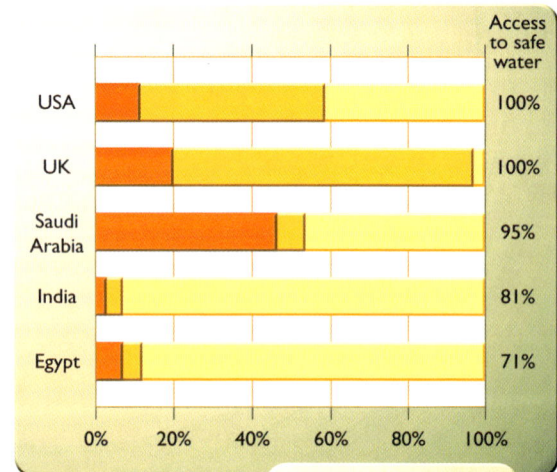

△ **Figure 6.9b** How water is used in some countries and how safe it is

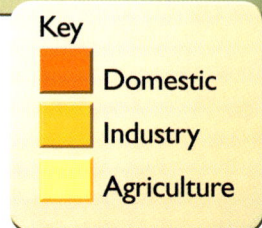

Country	Access to safe water
USA	100%
UK	100%
Saudi Arabia	95%
India	81%
Egypt	71%

Key
- Domestic
- Industry
- Agriculture

△ **Figure 6.10** Varanasi, a holy city on the banks of the Ganges

The Ganges River in India

The source of the River Ganges is melting snow from the Himalayas. **Monsoon rainfall** swells the river between June and September.

❯ The problem

The Ganges is a sacred river for Hindus. Six hundred kilometres of its 2 250 kilometres is heavily polluted. This pollution is caused by:

1 *Animal and human waste*. 40% of India's population live along the Ganges. Many cities do not have sewage works. Rainwater washes sewage into streams and rivers.
2 *Industrial waste*. Factories making paper, textiles and chemicals dump waste into the Ganges. Only 12 of the 132 factories treat their waste.
3 *Agriculture*. Fertilisers and pesticides wash from the fields into the Ganges.

❯ What can be done?

India is a Less Economically Developed Country (LEDC). It has few resources to cope with water pollution. More people are being born than dying. The country's population is growing rapidly. As the country's economy grows, more new factories are being built. A major problem is good water supply.
'It's difficult to get water of any kind, let alone clean water. And the problem can only worsen' (Indian pollution expert).

In 1985 The Ganges Action Plan began. It aimed to clean up the river. It cost over $300 million, with help from the United Nations. The project has built high technology sewage works. However, power cuts mean untreated sewage still goes into the Ganges. In Varanasi the pollution is 340 000 times the permissible level! The plan is a failure.

A local engineer says 'Our technology works. The plan is partly successful. We aim next to make the river safe for bathing.'

Amarnath, a pilgrim who bathes in the Ganges, ignores the rubbish, ash of **cremated** bodies and sewage. 'The river is pure. Germs cannot survive in its waters.'

› Is there a better solution to cleaning up the Ganges?

The head priest at Varanasi's Sankat Mochan temple is a scientist. He suggested a 12 km pipeline is used to take the sewage from Varanasi. The piped sewage would go into a series of ponds. Here **micro-organisms** would cleanse the water. The pipeline would use gravity to move the water. This would be simple and cheaper.

△ **Figure 6.11** Hindu pilgrims come to bathe in the Ganges. Hindus believe the waters will wash away their sins. Millions bathe each year in the Ganges

Activities

Varanasi's Mayor is worried about pollution in the Ganges. He wants the Ganges to be safe to use. He has three possible ideas:

A Leave the situation as it is. India needs the money to develop industry. As India becomes richer, it will be able to build a modern water and sewage system.

B Carry on with the Ganges Action Plan to build sewage works. The plan will take time to work. The Ganges is a big river system. Things will get better.

C Use micro-organisms in ponds to clear up sewage.

1 Copy and complete the table below.

Idea	Good point	Bad point
A Leave the situation		
B Ganges Action Plan		
C Use micro-organisms		

2 **a** Which idea would you choose?
b How will your idea help the people who live by the Ganges?
c Can you suggest another idea to clean up the Ganges? Draw and describe this.

ICT links

Check these links for more information about the Ganges.

www.csmonitor.com/durable/1997/10/29/intl/intl.6.html

http://ens.lycos.com/ens/jan2001/2001L-01-19-01.html

www.stfrancis.edu/ns/bromer/envhum/student8/

Are our oceans under threat?

The oceans

Oceans have most of the world's water. Their enormous size means that the threat of pollution is less than for rivers. In recent years, oil spills have caused major pollution in the oceans (Figure 6.12). Despite their large size, some oil spills caused little damage to the environment.

▽ **Figure 6.12** The worst oil spills in the world. Note: *Exxon Valdez* is ranked at 34

△ **Figure 6.13** The *Exxon Valdez* runs aground on Bligh Reef. Attempts to unload oil fail and large areas of the coast are polluted

Rank	Name of ship	Date	Location	Tonnes spilt
1	Atlantic Empress	1979	off Tobago, West Indies	287 000
2	ABT Summer	1991	700 nautical miles off Angola	260 000
3	Castillo de Bellver	1983	off Saldanha Bay, South Africa	252 000
4	Amoco Cadiz	1978	off Brittany, France	223 000
5	Haven	1991	Genoa, Italy	144 000
34	Exxon Valdez	1989	Prince William Sound, Alaska, USA	37 000

ICT Activities

1 Study Figure 6.12. On an outline world map plot these oil spills.

2 Check for other oil spills using this website: **www.itopf.com/stats.html**.

3 Scientists think that oil pollution has a long-term effect. A small rock affected by the *Exxon Valdez* spill has been photographed every year since that disaster.

To see the photos of this rock, log on to:

http://response.restoration.noaa.gov/photos/mearns/mearns.html

Write down three things you find out.

4 Search for more information about *Exxon Valdez*. http://projects.edtech.sandi.net/encanto/disaster/ shows you how to put together a project on the *Exxon Valdez* oil disaster.

How much land is damaged?

Building factories damages the land. The UK was the first country to become industrialised. The damaged and derelict land in the UK has been a problem.

Many former industrial areas have been **reclaimed**. Other industrial countries experience similar problems.

Reclaiming damaged land in South Wales

In recent years, many large factories in South Wales have closed. Steelworks at Cardiff, Ebbw Vale and Llanwern have all closed. Many coalmines have closed. This leads to a problem of what to do with vast areas of derelict land.

In Blaenavon, once a thriving coal and steel town, many shops have closed. The town has high unemployment and many empty buildings. Government grants have attracted some new industry. A mining museum named 'Big Pit' has brought some jobs.

△ **Figure 6.14** Llanwern steelworks before closure

◁▷ **Figure 6.15 and 6.16** New Tredegar – a mining village in 1958 (left) and today, as reclaimed land (right)

◁ **Figure 6.17** The site of Ebbw Vale steelworks became a **Garden Festival**, part of which houses a shopping centre

Activities

1 Study the photographs on page 112.
a Copy the concept map outline (Figure 6.18).
b Put at least three ideas for each part.

2 Can we have factories without the land being damaged? Suggest a reason for your answer.

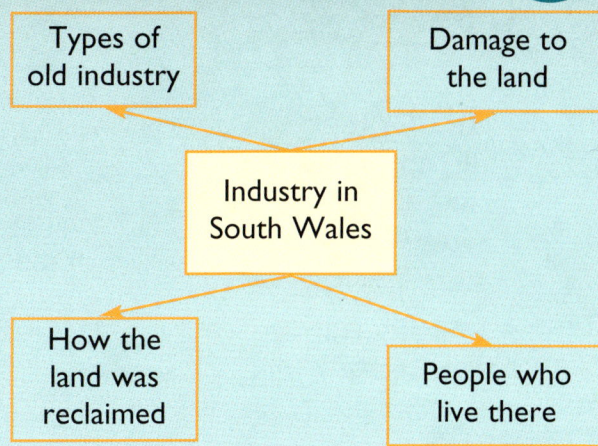

△ **Figure 6.18** Concept map

A success story from Schlema, Germany

Schlema is a town in Saxony, Germany. It once had serious problems of industrial pollution and derelict land. However, Schlema has overcome these difficulties. Try to piece together the story on the following pages.

Activities

1 Use the 12 clues on page 114 to find out why Hans Schmidt went back to Schlema.
a Copy each clue on a separate piece of paper.
b Sort the clues into piles. Some may go in more than one pile:
1 about Schlema; 2 about the Schmidts; 3 about rheumatism
c Make a list of new words that are in the clues. Use a dictionary to find out their meaning.

2 Prepare a talk to explain why Hans Schmidt went back to Schlema. Use these headings to help.
• How has Schlema changed since 1920?
• How did industry affect the people who lived in Schlema?
• How is Schlema helping ill people?

△ **Figure 6.19** Schlema's location in Germany

△ **Figure 6.20** Schlema's location in Saxony

In 1990 the mine was closed. The area was contaminated.

In 1947 the occupying Soviet forces began mining for uranium.

In the 1950s Hans returned to Schlema and was horrified by what he saw. He had relatives living in Schlema. He vowed that he would never go back.

The uranium mining destroyed the landscape.

In 2001 Hans was diagnosed as having rheumatism.

Schlema is a small town in Saxony, a region in Eastern Germany.

In 1992 the decision was taken to clean up the mine workings and reclaim the landscape.

In the 1920s Hans' father moved to Schlema. He suffered from rheumatism. The town's health **spa** was used to treat rheumatism.

The redevelopment will use the spring waters to treat rheumatism.

Schlema had iron, copper, cobalt and silver mines.

Hans was born and grew up in Schlema. He enjoyed living there and left during the 1939–45 war.

In the 1920s a radioactive spring was found. This was used to treat rheumatism.

▷ **Figure 6.21** Hans Schmidt and family

△ **Figure 6.22** Reclamation in Schlema.
Waste rock pile: 1960, 1993, 1993

> **Have you solved the mystery?**

• Do you know why Hans Schmidt is going back to Schlema?
• Would you go back?

ICT links

Websites to take you further:
http://forests.org/articles/reader.asp?link id=8247
An article on how Germans are planting trees to reclaim the land.
http://abcnews.go.com/sections/world/ DailyNews/Radioactivespa010118.html
American report on the benefits of radiation at Schlema.
www.bmwi.de/textonly/Homepage/down load/english/wismut_e.pdf
This is an English language account of land reclamation in Saxony.

△ **Figure 6.23** The uranium mine, now a tourist attraction

What a choke!

Air pollution is when harmful gases are found in the air. Using fossil fuels causes air pollution. Cars, power stations and industry all use fossil fuels and produce air pollution.

The car's the star – or is it?

We use cars everyday, but cars cause air pollution. Los Angeles, a large US city, has very high car ownership. The city lies in a basin. The basin traps the air pollution (Figure 6.24). Attempts to reduce pollution by fitting cars with **catalytic converters** have had little effect.

△ **Figure 6.24** A photo from Los Angeles showing air pollution!

As the population grows and more people can afford a car, traffic problems will increase. More vehicles cause traffic jams (Figure 6.25). Pollution from the vehicles covers the area. In 1970, Mexico City had a new subway system built. The city also banned using cars on certain days of the week. However, the traffic problems are still bad.

▽ **Figure 6.25** Traffic in the streets of Mexico City

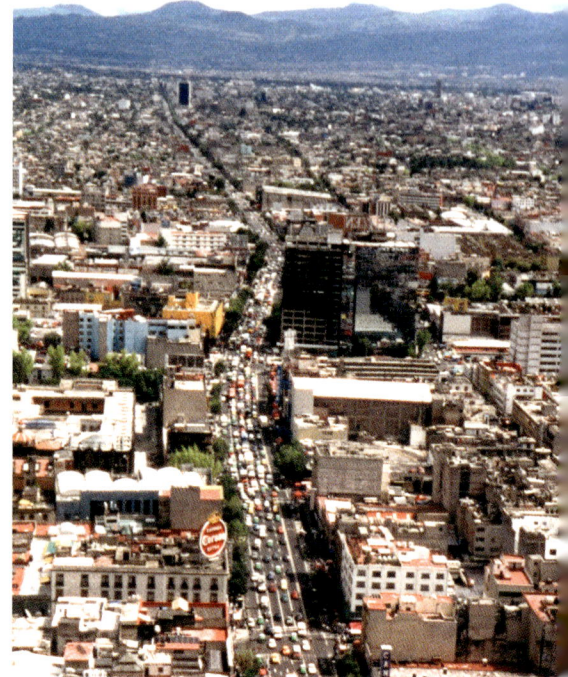

	Number of cars per thousand population	Predicted population in millions 2025
USA	521	333
UK	359	60
Mexico	92	130
India	4	1330
China	3	1480

△ **Figure 6.26** Tables of number of cars/predicted population

Activities

1 Describe two problems caused by traffic jams.

2 Suggest two ways to reduce these problems.

3 Study Figure 6.26. Which countries are likely to have car pollution problems?

4 Draw a cartoon to show the problems if everyone owned a car.

Does air pollution mean wealth?

Medieval London had air pollution. More Londoners used coal for their domestic fires. This meant London's air was unhealthy.

The demand for electricity has seen bigger power stations built (Figure 6.27). The coal burnt damages the atmosphere. New houses built near the power stations suffer from pollution (Figure 6.28).

Many developing countries ignore air pollution in the rush to get rich. India is the sixth largest producer of **greenhouse gases**. The Chinese city of Benxi is an important centre for steel and cement production. Seven million tonnes of poor quality coal, high in sulphur, are used for electricity, industry and in the home. This pollutes the air and the government does little to prevent it. A plan to reduce pollution from the steelworks and use the waste gas to heat 50 000 houses is in place.

However, air pollution will cause 2.5 million worldwide deaths.

◁ **Figure 6.27** Pollution from electricity generation, Czech Republic

◁ **Figure 6.28** Demand for new housing despite being close to a polluter

◁ **Figure 6.29** Heavy industrial pollution in the Liaoning **Province** of north-east China

Common air pollutants	
Smoke	This is easily seen as it is made up of small solids.
Sulphur dioxide	This is a colourless gas mainly coming out of **thermal power stations.** When mixed with water, **acid rain** is produced.
Carbon monoxide	This gas has no colour or smell. It is produced by road transport.
Hydrocarbons	Produced when petrol is not fully burnt. Helps the formation of **smog**.
Nitrogen oxide	Given out by vehicles and power stations. High amounts in cities.
Ozone	Main gas in **photochemical smog** produced when nitrogen oxides and hydrocarbons react in sunlight.
Particulates	Small bits of solid or liquid, e.g. soot, fumes carried in the air. Produced by vehicles, industry and domestic fires.

△ **Figure 6.30** Table of common air pollutants

Activities

5 **a** Would you like to live next to a polluting factory?
b Suggest a reason for your answer.

6 Study Figure 6.30. Design a poster to help people understand air pollution.

Should we think globally, act locally?

Many environmental problems are global. If they are global, how can we help? We can do some things as individuals and others must be done by local authorities.

In 1992, United Nations held a conference called the Earth Summit in Rio de Janeiro. Over 150 countries attended, including Britain, and they all signed up for **Agenda 21**. This sets out how countries can work together to create sustainable development. Agenda 21 deals with climate change and **deforestation**. It calls for local people and councils to work together to develop plans for a more sustainable lifestyle.

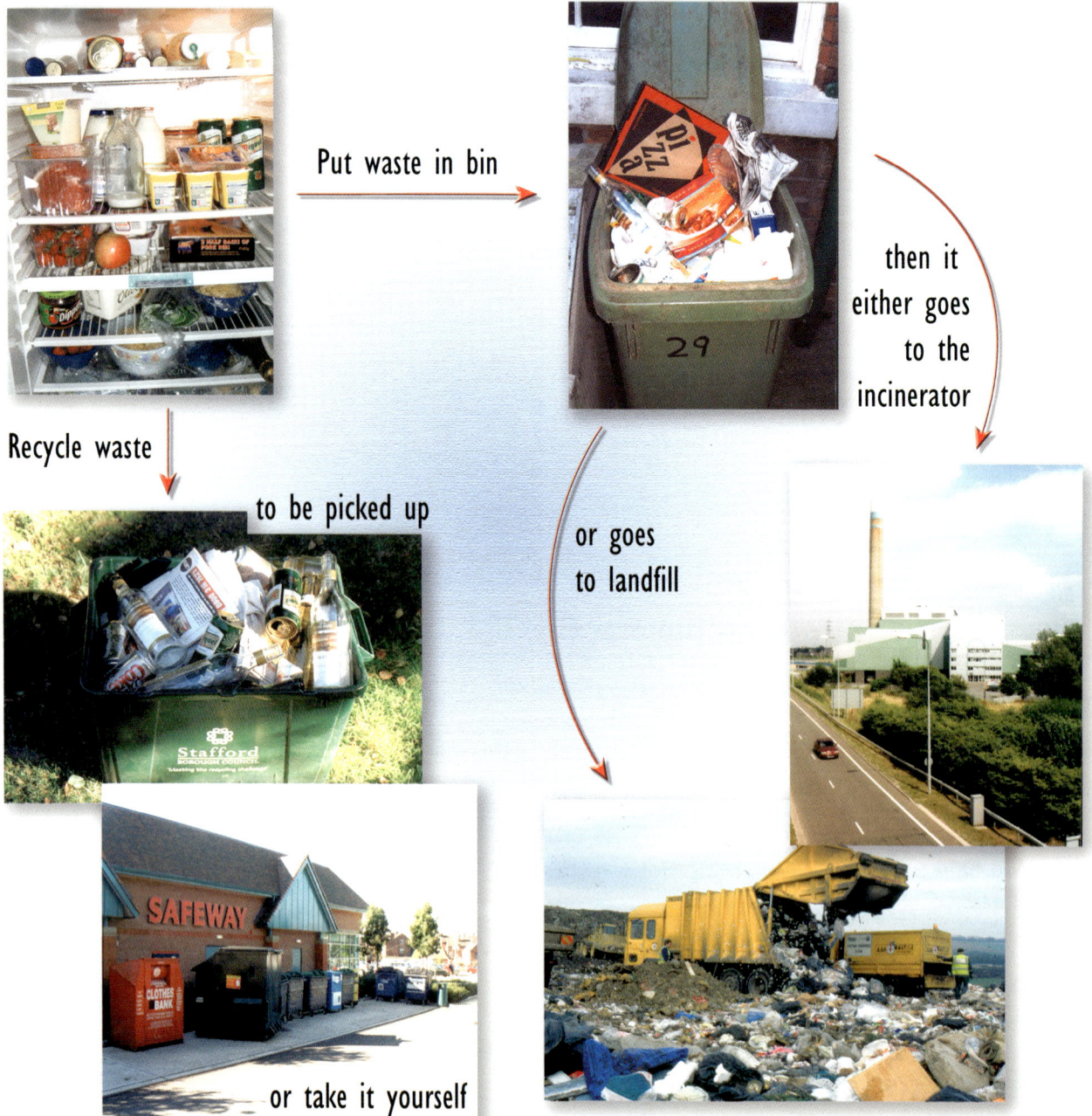

Put waste in bin

then it either goes to the incinerator

Recycle waste

to be picked up

or goes to landfill

or take it yourself

△ **Figure 6.31** Here are some of the possible routes of household waste

▷ **Figure 6.32** Recycling green waste

Activities

Figure 6.31 shows you where your waste might end up.
Your local council has called a meeting to see how far they have got with Agenda 21.
You are a reporter for a local newspaper. You want to write a story about this issue. Your feature should fill one page of a newspaper.

It should have:
a an eye-catching headline
b an introduction to explain what Agenda 21 is
c some ideas about what can be done
d some good points
e some problems
f a summing up

Milk bottle recycling competition
Win a portable TV and playstation

There are 5 prize packs to give away including a £50 shopping voucher and a T-shirt made from recycled plastic. One lucky person will also win a portable colour television and a playstation!!

To enter, simply write your name and telephone number / contact details on each milk bottle, remove the caps, squash the bottles and put them in this bank.

This competition is run by Recoup. Full competition details and lots more information about plastic bottle recycling is available on our website at www.recoup.org

△ **Figure 6.33** Can competitions encourage more recycling?

119

Tasks

The *Khian Sea* was a 'garbage ship' that spent years looking for somewhere to dump its waste. Study Figure 6.34.

1 What waste were they trying to get rid of?

2 How many countries did the *Khian Sea* try to unload its cargo onto?

3 How did the *Khian Sea* get rid of its load?

4 Do you think rich countries should dump their waste on poorer countries?

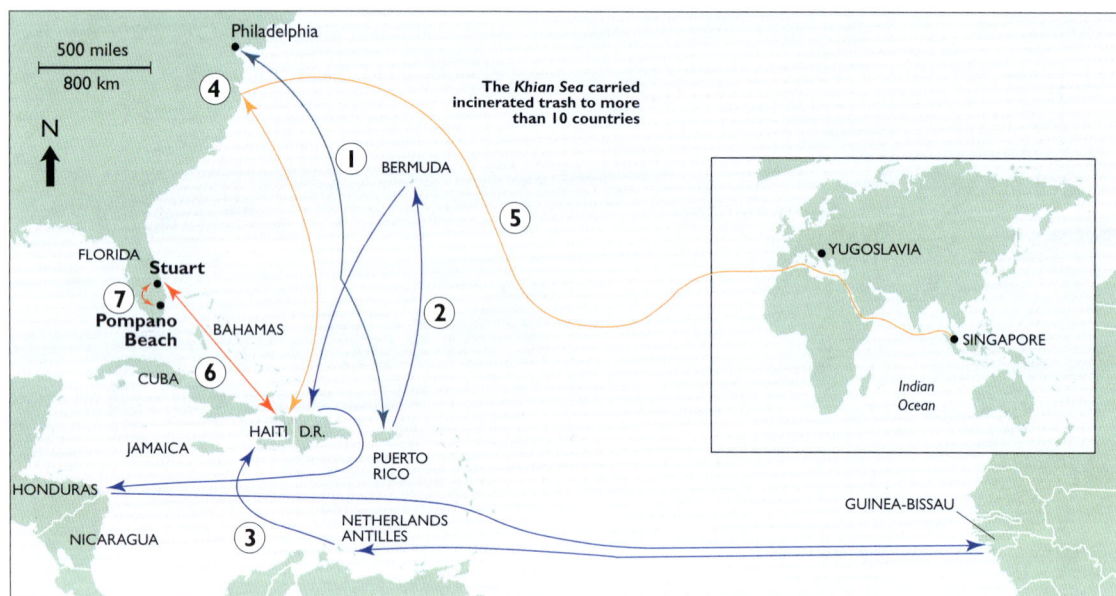

① **Sept. 5 1986**
The *Khian Sea* leaves Philadelphia for the Bahamas. It has 14 355 tons of ash. The Bahamian government refuses to allow the ship to dock. The *Khian Sea* has to change course.

② **Sept. 1986–Aug. 1987**
The *Khian Sea* is refused entry to Puerto Rico, Bermuda, the Dominican Republic, Honduras, Guinea-Bissau and the Netherlands Antilles.

③ **Dec. 1987–Feb. 1988**
The *Khian Sea* docks and unloads 4 000 tons of ash in Haiti. The government stops the unloading. The *Khian Sea* is ordered to leave.

④ **March 1, 1988**
The *Khian Sea* arrives in Delaware Bay. The ship is not allowed to unload. The ship leaves for Africa.

⑤ **Aug.–Nov. 1988**
The ship arrives in Yugoslavia but has changed its name to *Felicio*. It arrives empty in Singapore with a third name, *Pelcano*. The captain admits he dumped the ash in the Atlantic and Indian Oceans.

⑥ **Spring 2000**
The ash dumped in Haiti is picked up and taken to the USA.

⑦ **Thursday, 26 January 2001**
The ash is buried in a landfill site near Pompano Beach.

△ **Figure 6.34** The story of the *Khian Sea*. Organisations like Basel Action Network try to stop rich countries dumping their waste in poorer countries. See their website **www.ban.org**

Tasks

We need to use our land carefully. You may know about a local issue concerning **greenfield** sites. Study the photographs on this page. Some show a greenfield site and others a **brownfield** site.

5 Which two photographs are brownfield sites? How was the land used before?

6 Suggest one advantage and one disadvantage of using brownfield sites for building new houses.

7 Which two photographs show greenfield sites? How was the land used before?

8 Suggest one advantage and one disadvantage of using greenfield sites for building new homes.

9 Why is Figure 6.37 different from the other photographs?

10 Suggest an idea that would help improve your local area.

△ **Figure 6.35** Cows graze on a greenfield site, but not for long

△ **Figure 6.36** New houses built on farmland (a greenfield site). The **rural–urban fringe** is under threat

◁ **Figure 6.37** An old shoe factory built in Stone, Staffordshire. Today its location, size and shape are less attractive to industrialists

▽ **Figure 6.38** The factory is demolished and a number of 'town houses' are built on this brownfield site

▽ **Figure 6.39** The site is cleared

Assessment tasks

▽ **Figure 6.40** What London uses and wastes

Uses	Tonnes per year
Fuel (oil equivalent)	20 000 000
Oxygen	40 000 000
Water	1 002 000 000
Food	2 400 000
Timber	1 200 000
Paper	2 200 000
Plastics	2 100 000
Glass	360 000
Cement	1 940 000
Bricks, blocks, sand, tarmax	6 000 000
Metals	1 200 000

Waste	Tonnes per year
Carbon dioxide	60 000 000
Sulphur dioxide	400 000
Nitrogen oxides	280 000
Sewage	7 500 000
Industrial waste	11 400 000
Household	3 900 000

1. Input this data (Figure 6.40) on an Excel spreadsheet.

2. Use graph wizard to present this information.

3. Export your graphs into Publisher to produce a leaflet about how London uses and wastes resources.

4. Write a comment to describe your graph.

5. Think of a catchy title for your leaflet.

6. Design a poster that highlights the problems of London's use of resources. Use this statement:
   'Cities occupy 2% of the land surface of the world but use 75% of the world's resources.'
   Your poster should try to suggest a solution.

△ **Figure 6.41** Large cities such as London create large quantities of industrial waste

Checking out websites

There are many websites on environmental issues. For each website below, say what you liked about it, what you disliked about it and give it a score out of ten.

Website Warning!

Try to think why websites are there. Who pays for them? Do they give all points of view? Are they truthful?

www.wastewatch.org.uk/
Go to the Kids Home Page

www.foe.org.uk/campaigns/industry_and_pollution/factorywatch/
Find out about pollution near you. Just type in your postcode, and a map with a list of polluters will be shown.

www.doingyourbit.org.uk
How you can help the environment.

www.globalactionplan.org.uk
Small changes can make a difference.

www.aeat.co.uk/netcen/airqual
Information on air quality in Britain. Find the education section.

ICT Activity

Some websites can give you details about your local area, for example, factories. You search using your postcode details.

Figure 6.42 is from the **www.foe.org.uk/campaigns/industry_and_pollution** site. It shows the Liverpool area. This is an industrial area with many chemical plants along the Mersey Estuary. One plant is an oil refinery belonging to Shell. Its postcode is L65 4HB. By typing this postcode into 'Factory Watch' on the website above, details of chemicals put into the air will be listed.

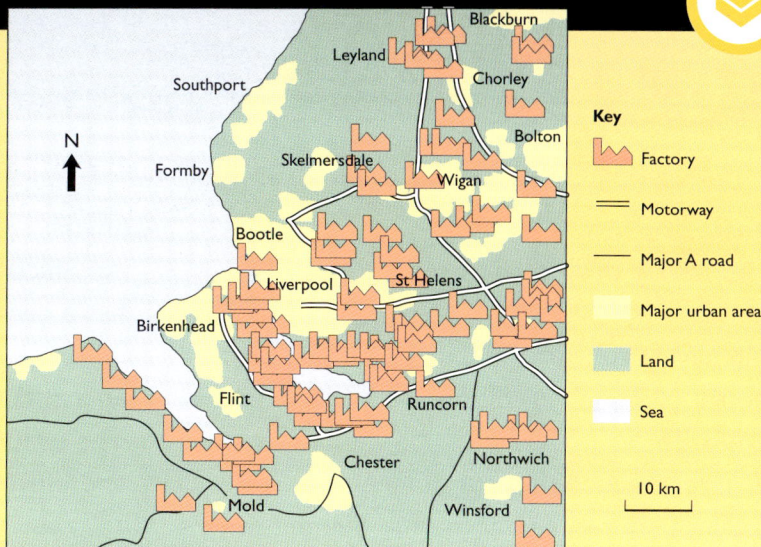

△ **Figure 6.42** Map of the Liverpool area of north-west England, showing chemical plants

Friends of the Earth

◁ **Figure 6.43** The logo of **Friends of the Earth**

Review

Environment Quiz

Choose the correct answer:

1 Sustainable development is:
   a using today with tomorrow in mind
   b taking as much from the Earth as possible, regardless of the consequences

2 Agenda 21:
   a is an idea to raise the drinking age to 21
   b was agreed at the Earth Summit in 1992. It encourages local action on sustainability

3 Global warming is mainly caused by:
   a producing greenhouse gases
   b nuclear power stations

4 Which are not examples of renewable energy?
   a coal, oil, natural gas
   b wind, solar, tidal power

5 LEDC countries are becoming more polluted because:
   a population is falling and industry is closing down
   b population is rising and there is more industry

6 Schlema is:
   a a town in Saxony where land has been reclaimed
   b a device to control factory emissions

7 Air pollution is:
   a caused by emissions from cars
   b caused by barbecues

8 Landfill is used to:
   a dispose of waste
   b generate electricity

9 Recycling aims to:
   a re-use waste materials
   b build cycle lanes on busy roads

10 A brownfield site has:
   a been ploughed and so looks this colour
   b has been built on before

If you're not really sure – the answers are on page 129!

Glossary

Abiotic

A non-living thing such as soil

Acid rain

Rain having sulphuric and nitric acid strong enough to harm vegetation and freshwater life

Agenda 21

An agreement about environmental sustainability reached at the Rio Summit in 1992

Amenities

Something that benefits the residents of an area. This could include a swimming pool or a good shopping area

Ancestors

People from whom your family is descended

Annotate

 Add notes to explain

Biomes

Global scale ecosystems

Biotic

A living thing, such as a rabbit

Birth rate

The number of births per thousand people in a year

Brownfield

An area for development that has already been built on

Car pool

The sharing of cars in journeys to work

Carnivore

An animal which eats meat

Cartographers

People who design and draw maps

Catalytic converter

A device fitted to a car exhaust to reduce the emission of poisonous gases

Census

An official count of population data. The UK has a census every ten years, e.g. 1991, 2001, 2011

Central Business District (CBD)

The centre of towns and cities where shops and offices are found

Civil war

A war between people from the same country

Components

Interconnected parts of an ecosystem

Conflict

A struggle between people when they have different ideas about how to use the same resources

Constructive plate margin

Where new crust is formed at a mid-ocean ridge

Consumer

Something which eats something else. In a food chain, the creature which eats the plants is referred to as a primary consumer; the creature which eats the primary consumer is called the secondary consumer and so on.

Counter-urbanisation

The movement of people from housing in towns and cities to the countryside

Cremation

The burning to ashes, usually of human remains

Death rate

The number of deaths per thousand people in a year

Decomposer

An organism which lives off the remains of dead plants and animals, speeding up the process of decay, e.g. bacteria or fungi

Deforestation

Cutting trees down without replanting

Democracy

Government by elected representatives of the people

Dependent population

Total population under 15 and over 65

Derelict

Unused and abandoned land

Destructive plate margin

Where crust is destroyed as it is taken back down into the mantle

Dictator

A person with complete authority

Digital divide

The gap between people and countries who have access to technology and those who don't

Directory

A book published in the 19th- and early 20th centuries listing the residents of an area and their occupations

Distribution

The spread of something throughout an area

Energy

Force needed to do work – can be renewable or non-renewable

Entrepreneurship

The ability to develop a new idea, particularly as a business, often at personal financial risk

Environment

Our surroundings – natural environment refers to land, water and air

Epicentre

The point on the Earth's surface directly above the focus.
The centre of the earthquake on the surface.

Eutrophication

What happens when nutrients from fertilisers are washed into areas of freshwater, causing plants like algae to grow rapidly. As they grow, all the oxygen in the water is used up, leaving none for other organisms in the ecosystem. The result can be dead and lifeless lakes and rivers

Evolve

To grow and adapt to an ecosystem

Extinction

Complete destruction of animal and plant species

Faults

Weaknesses in rocks which can be easily affected by earthquakes

Focus

The point within the Earth where the rocks slip causing the earthquake.

Food chain

A sequence of living things which feed off each other, e.g. a wolf eats a reindeer which eats grass

Food web

A set of interconnected food chains, such as a food web of all living things in a hedgerow

Fossil fuels

Created in past geological periods from hydrocarbons, e.g. coal, natural gas and oil

Friends of the Earth

A non-governmental organisation of people who support the environment and conservation

Garden Festival

An attempt to reclaim derelict land in Britain from old industrial sites

Global

Worldwide or whole world distributions, e.g. earthquakes, population densities

Global warming

The heating up of the Earth's atmosphere

Globalisation

Companies operating worldwide, e.g. Coca Cola

Greenfield

A site that has not been built on before

Greenhouse gases

Gases which contribute to global warming such as carbon dioxide, methane, ozone

Hamlet

A small rural settlement without services

Herbivore

An animal which eats plants

Hotspots

Weaknesses in the Earth's crust which allow **magma** to erupt from the mantle

Immigrant

A person who moves into a country permanently

Incineration

Burning to ashes

Individualism

Free action by any one person

Informal job

One which is 'unofficial' such as shoe-shining, selling food from street stalls

International migrants

People who move from one country to another. Some move for a better life, others are forced

Landfill

A site used to bury waste, usually a hole in the ground

Lava

The name given to molten rock when it erupts from a volcano

LEDC

Less Economically Developed Country

Life expectancy

The average number of years someone is expected to live

Magma

The name for the molten rock of the mantle

Mantle

The part of the Earth, below the crust, that is liquid, molten rock

MEDC

A More Economically Developed Country

Mercalli Scale

A scale describing the effects of earthquakes

Micro-organism

A tiny animal or plant

Migrate

A general term to describe the movement of population. This can be either permanent or temporary

Monsoon rainfall

A period of heavy summer rainfall

Nation

People having a common language, history etc.; overseen by one government

Natural hazard

Physical processes or events that have the potential to cause harm to life or property

Non-renewable

Something that cannot be made again and is finite, e.g. fossil fuels

Open source software
When the thinking behind a program is freely available to all

Monsoon rainfall
A period of heavy summer rainfall

Pattern
The way in which features occur or are arranged

Peninsular
Piece of land almost surrounded by water

Photochemical smog
Created when exhaust fumes and factory emissions combine with sunlight to produce ozone

Plateau
A large fairly flat area of upland

Pollution
When something is contaminated by human activity

Population pyramid
A graph showing the age and sex of a country's population

Producer
Plant which makes (produces) energy by photosynthesis

Province
An administrative division of a country

Pull factors
Reasons why people are attracted to other places and move there, e.g. to get a better job

Push factors
Reasons why people leave places, often from villages, to live in towns and cities, e.g. through lack of land

Raw materials
Anything that can be processed to make a product, e.g. iron ore is one of the raw materials to make steel

Reclaimed land
Either damaged land restored to a useful state again or land converted to agriculture from the sea, marsh or desert

Recycling
Re-using materials which would normally be thrown away, e.g. paper, drink cans

Region
An area that may be made up of part of one or several countries, that has either physical or human feature(s) unique to it

Renewable
Resources which can be used repeatedly such as hydro-electric power (HEP), wind, wood

Residential
Areas of housing

Richter Scale
A scale for measuring the force of earthquakes

Rural
The countryside

Rural–urban migrant
A person who moves from the country to a town or city

Rural–urban fringe
The zone on the edge of a urban area (town or city) which comes into contact with a rural (country) area

Settlement
Any form of habitation from a single house to a large city

Shanty town
In a city of a developing country where houses are built illegally in a makeshift way

Shield volcano
Flat volcano that produces lava

Smog
A mixture of smoke and fog

Spa

A place where mineral springs are found, named after Spa in Belgium

Strato volcano

Usual volcano shape that produces lava and ash.

Sustainable development

To use the resources of an ecosytem in a way that ensures they are replaced for future generations.

Technology

Scientific, industrial and mechanical innovations

Tectonic activity

Events caused by the movement of the Earth's surface, e.g. earthquakes and volcanoes

Tectonic plates

Parts of the Earth's crust that float on top of the liquid mantle

Telework

To work at home using telephones and computers to contact others

Thermal power station

Makes electricity by heating water into steam which turns generators

Tsunami

A large sea wave generated in the oceans by earthquake or landslides

United Nations

An organisation of most countries in the world. It tries to promote peace, security and international co-operation

Urban

Of towns and cities

Vegetation

Plants

Village

A small rural settlement larger than a hamlet often with a church, shop and public house

Vulcanologist

A person who studies volcanoes

(Answers to Environment Quiz on page 123: 1a; 2b; 3a; 4a; 5b; 6a; 7a; 8a; 9a; 10b)

Index

CL

910
ARU